WIT Libraries Luke Wadding Library

Long term loan (Overdues charged at 50c a day)

Luke Wadding Library, Waterford Institute of Technology,
Cork Road, Waterford, Ireland
Tel: +353.51302823 email: libinfo@wit.ie 🅣 @witlibraries

VIDEO AND IMAGE PROCESSING IN MULTIMEDIA SYSTEMS

THE KLUWER INTERNATIONAL SERIES
IN ENGINEERING AND COMPUTER SCIENCE

MULTIMEDIA SYSTEMS AND APPLICATIONS

Consulting Editor

Borko Furht
Florida Atlantic University

VIDEO AND IMAGE PROCESSING IN MULTIMEDIA SYSTEMS

by

Borko Furht
Florida Atlantic University
Boca Raton, Florida, USA

Stephen W. Smoliar
HongJiang Zhang
Institute of Systems Science
National University of Singapore

KLUWER ACADEMIC PUBLISHERS
Boston / Dordrecht / London

Distributors for North America:
Kluwer Academic Publishers
101 Philip Drive
Assinippi Park
Norwell, Massachusetts 02061 USA

Distributors for all other countries:
Kluwer Academic Publishers Group
Distribution Centre
Post Office Box 322
3300 AH Dordrecht, THE NETHERLANDS

Library of Congress Cataloging-in-Publication Data

Furht, Borivje.
 Video and image processing in multimedia systems / Borko Furht,
Stephen W. Smoliar, HongJiang Zhang.
 p.   cm. -- (The Kluwer international series in engineering and
computer science)
 Includes bibliographical references and index.
 ISBN 0-7923-9604-9 (alk. paper)
 1. Multimedia systems. 2. Video compression. 3. Image
processing--Digital techniques. I. Smoliar, Stpehn W. II. Zhang.
HongJiang. III. Title. IV. Series.
QA76.575.F87   1995
006.6--dc20   95-30423
                                                             CIP

Printed on acid-free paper.

Printed in the United States of America

Dedication

To Sandra, Linda, and XueFeng

with our heart-felt thanks for all their understanding and patience.

CONTENTS

Acknowledgments

Many of the images in this book have been reproduced with the kind permission of many publishers and other sources. We would like to thank to the IEEE for allowing us to reproduce Figures 2.11, and 2.13-17 from *IEEE MultiMedia* and Figures 1.12-13, 3.6-8, and 3.10-13 from *IEEE Computer*. We should acknowledge that many figures in Chapters 5,6 and 7 will be appearing in two survey papers on multimedia compression techniques and standards published in the *Journal of Real-Time Imaging* (Academic Press, 1995). Figures 8.5-6 were reproduced from *Byte*, and Figures 8.7-10 were reproduced with permission of Kluwer Academic Publishers.

We are particularly thankful to a number of companies for providing information on their products and systems, and allowing us to use their figures and results in Chapters 9 and 10. These companies include: AT&T, Autograph International, Hewlett-Packard, IBM, Integrated Information Technology, Intel, LSI Logic, New Media Graphics, Optivision, Siemens, Sun, Production Catalysts, Telephoto, and Texas Instruments.

We are also particularly indebted to Springer-Verlag for allowing us to reproduce the following Figures which have appeared in issues of *Multimedia Systems*: Figures 12.1-6, Figures 14.1-7, and Figures 14.9-10. In addition, Figure 14.8 will be appearing in the Springer-Verlag volume *Advances in Digital Libraries*. Similarly, all the Figures in Chapter 13 have been taken from the Kluwer journal *Multimedia Tools and Applications*; and Figure 14.11 has been reproduced with permission from the IEEE. The Cunningham Dance Foundation has been particularly generous allowing their material to be used as test data; and Figures 12.7, 12.10, and 12.11 have been reproduced with their permission. Finally, Figure 11.9 was reproduced with the permission of the Television Corporation of Singapore (formerly the Singapore Broadcasting Corporation).

The results reported in Part II could not be produced without contributions from many graduate students working in the Multimedia Laboratory at Florida Atlantic University (FAU), funded by US National Science Foundation. Ray Westwater, Mark Kessler, Jan Alexander, Mauricio Cuervo, Lou Horowitz, and Henry Pensulina have developed the sequential JPEG algorithm and provided results and images presented in Part II. Shen Huang and Zhong-Gang Li have performed and provided results for progressive JPEG compression, while Peter Monnes has analyzed JPEG performance. Srikar Chitturi has ported and analyzed performance of the Berkeley software MPEG decoder. Keith Morea has implemented fractal image compression and provided related results, while Pavani Chilamakuri and Dana Hawthorne have evaluated subband coding versus JPEG and MPEG techniques and have provided those results.

The research reported in Part III could not have been carried out without the support   granted by the National Science and Technology Board of Singapore to the Institute of Systems Science (ISS) We also wish to thank Juzar Motiwalla, Director of ISS, for all the encouragement he has given to the members of the Video Classification Project during the time this work was conducted. The researchers at ISS are also particularly indebted to Professor Louis Pau, who has served in an advisory capacity since the commencement of this effort and has provided no end of helpful suggestions. In addition the authors wish to acknowledge the many contributing members of the ISS team, including Chien Yong Low, Jian Hua Wu, Atreyi Kankanhalli, You Hong Tan, and Siew Lian Koh.

Special thanks for typing, formatting, and finalizing the book goes to Donna Rubinoff from FAU, whose expertise and dedication were invaluable in the completion of this work.

PART I
Introduction to Multimedia

1

BASIC CONCEPTS

1.1 DEFINITION OF MULTIMEDIA

Multimedia computing has emerged in the last few years as a major area of research. Multimedia computer systems have opened a wide range of potential applications by combining a variety of information sources, such as voice, graphics, animation, images, audio, and full-motion video. Looking at the big picture, multimedia can be viewed as the merging of three industries: computer, communication, and broadcasting industries.

Research and development efforts in multimedia computing have been divided into two areas. As the first area of research, much effort has been centered on the stand-alone multimedia workstation and associated software systems and tools, such as music composition, computer-aided learning, and interactive video. However, the combination of multimedia computing with distributed systems offers even greater potential. Potential new applications based on distributed multimedia systems include multimedia information systems, collaboration and conferencing systems, on-demand multimedia services, and distance learning.

1.1.1 Multimedia Elements

The fundamental characteristic of multimedia systems is that they incorporate continuous media, such as voice, video, and animated graphics. This implies the need for multimedia systems to handle data with strict timing requirements and at high rate. The use of continuous media in distributed systems implies the need for continuous data transfer over relatively long periods of time (e.g.

playout of video stream from a remote camera). Additional important fundamental issues are: media synchronization, very large storage requirements, and
need for special indexing and retrieval techniques, tuned to multimedia data
types [Fur94].

The fundamental media, graphics, animation, sound, and motion, including full
color, are briefly discussed next.

Color has been a part of computers for a long period of time. Colors are created
through a combination of three primary colors, red, green, and blue (RGB). For
more details on RGB and other color formats see Section 4.3. During the last
decade, the resolution and the number of colors displayed on a computer screen
have increased. Table 1.1 shows the most common display modes supported by
current multimedia systems.

Graphic	Image Resolution	Colors
VGA	640x480	16
VGA8	360x480	256
BGA	640x480	256
XGA	1024x768	256

Table 1.1   Display modes supported by current multimedia systems.

The standard display mode is VGA (Video Graphics Array), which provides
a resolution of 360x480 pixels and 16 colors. The VGA-8 supports the same
resolution of 360x480 pixels, however the number of colors is 256. The Extended
Graphics Array (XGA) display mode provides a resolution of up to 1024x768
pixels and 256 colors.

There is a tradeoff between the resolution and the number of colors. Most
application developers prefer a larger amount of colors to higher resolution,
because human eyes are more sensitive to colors than resolution.

Graphics. With the evolution of advanced high-quality display modes, which support more colors and higher resolution, the computer graphics and images have been significantly improved and are part of many multimedia applications.

Animation is a series of images, each slightly different from the previous, displayed in successive time units so they can appear as moving pictures. The typical frame rate required for animation is 15-19 frames per second.

Sound and motion. The ultimate characteristic of multimedia systems is the incorporation of sound and motion video. Motion video requires a powerful workstation or a personal computer, a high-quality display mode, a large amount of storage capacity, and the usage of compression techniques. A typical frame rate required for motion video is 24 frames/second for movies, 25 frames/second for PAL systems, and 30 frames/second for NTSC systems.

Figure 1.1 illustrates basic principles in dealing with continuous media (audio and video) and still images, as well as several examples of operations on these media. Audio and video information can be either stored and then used in an application, such as training, or can be transmitted live in real-time. Live audio and video can be used interactively, such as in multimedia conferencing, or non-interactively, in TV broadcast applications. Similarly, still images, which are stored, can be used in an interactive mode (using operations such as browsing and retrieval) , or in non-interactive mode (slide show).

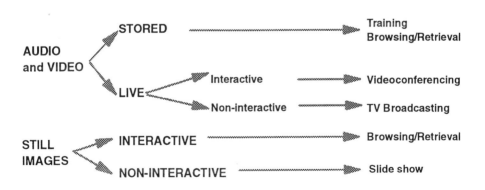

Figure 1.1 Various modes of operation on multimedia data.

1.1.2   Computer Systems and Multimedia

It is interesting to note that multimedia stresses all the components of a computer system, as illustrated in Figure 1.2. At the processor level, multimedia requires very high processing power in order to implement software codecs, and multimedia file systems and corresponding file formats. At the architecture level, new solutions are needed to provide high bus bandwidth, and efficient I/O. At the operating systems level, there is a need for high-performance real-time multimedia OS, which will support new data types, real-time scheduling, and fast interrupt processing. Storage and memory requirements include very high capacity, fast access times, and high transfer rates. At the network level, new networks and protocols are needed to provide high bandwidth and low latency and jitter required for multimedia. At the software tool level, there is a need for new object-oriented, user-friendly multimedia development tools, and tools for retrieval and data management, important especially for large, heterogeneous, networked and distributed multimedia systems.

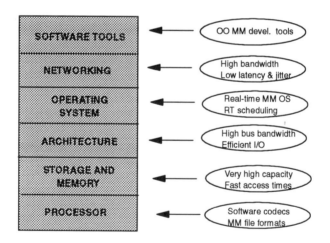

Figure 1.2   Multimedia stresses all the components of a computer system.

Researchers in the multimedia field are currently working in existing computer areas with the goals of transforming existing technologies or developing new ones suitable for multimedia. These research areas include most of the areas in the computer field such as: fast processors, high-speed networks, large-capacity storage devices, new algorithms and data structures, video and audio compression algorithms, graphics systems, human-computer interaction, real-time operating systems, object-oriented programming, information storage and retrieval, hypertext and hypermedia, languages for scripting, parallel processing methods, and complex architectures for distributed systems [Fur94, FM95].

1.2 MULTIMEDIA WORKSTATIONS

A multimedia workstation can be defined as a computer system capable of handling a wide variety of information formats: text, voice, graphics, image, audio, and full-motion video. Advances in several technologies are making multimedia systems technically and economically feasible. These technologies include powerful workstations, high capacity storage devices, high-speed networks, advances in image and video processing (such as animation, graphics, still and full-motion video compression algorithms), advances in audio processing (such as music synthesis, compression, and sound effects), and speech processing (speaker recognition, text-to-speech conversion, and compression algorithms).

Since 1989, when the first multimedia systems were developed, it is already possible to differentiate three generations of multimedia systems, as presented in Table 1.2 [Col93].

The first generation, based on i80386 and MC68030 processors, was characterized by bit-mapped image and animation as media used, JPEG as the video compression technique applied, and local area networks based on Ethernet and token rings. Authoring tools were based on hypermedia. The second generation of multimedia systems used i80486 and MC68040 processors as computer platforms, the media included moving and still images and 16-bit audio, video compression technology was based on moving JPEG and MPEG-1, and the local area network was 100 Mb/s FDDI . The authoring tools were based on object-oriented multimedia, which included text, graphics, animation, and sound.

We are presently at the transition stage from second to third generation multimedia systems, based on more powerful processors, such as Pentium and PowerPC. Media used are full-motion video of VCR quality, and in the future NTSC/PAL and HDTV quality. The compression algorithms use MPEG-2, and perhaps the wavelets that are now in the research stage. Faster and enhanced Ethernet , token ring , and FDDI networks are used, as well as new

	FIRST GENERATION 1889-91	SECOND GENERATION 1992-93	THIRD GENERATION 1994-95
MEDIA	Text, B/W graphics Bit-mapped images Animation	Color bit-mapped images 16-bit audio Moving still images Full motion video (15 f/s)	Full-motion video (30 f/s) NTSC/PAL & HDTV quality
AUTHORING CAPABILITY	Hypertext Hypermedia	Object-oriented MM tools with text, graphics, sound, animation, images, and motion video.	Integration of object-oriented MM tools with OS
VIDEO COMPRESSION TECHNOLOGY	DCT JPEG	Motion JPEG MPEG-1 Px64	MPEG-2,4 Wavelets
OPERATING SYSTEM	DOS	DOS 5, OS/2 Windows 3.X UNIX, MAC OS	Windows NT Warp Chicago Pink (IBM/Apple)
BASE PLATFORM	15 MHz 386 (68020) 2 MB DRAM 40 MB hard disk VGA color (680x740) 500 MB CD-ROM (100 KBps)	50 MHz 486 (68040) 8-16 MB DRAM 240-600 MB hard disk 1-2 1.5 MB floppies VGA with 256 colors (1024x768) 500 MB CD-ROM (150 KBps)	50-100 MHz Pentium (PowerPC) 16-32 MB DRAM 1-2 1 GB hard disk 20-30 MB floppies SVGA (1280x960) 600 MB CD-R (300 KBps)
DELIVERY MODE	720 KB diskette 1.5 MB laser disk (R/O) 128 MB CD-ROM (R/O)	500 MB CD-ROM (R/O)	600 MB CD-R 500 MB WORM 128-500 MB magnetooptic (R/W)
LOCAL-AREA NETWORK	Ethernet (10 Mbps) Token ring (16 Mbps)	FDDI (100 Mbps) Priority token ring Switched Ethernet	Ethernet, Token ring (100 Mbps) FDDI (500 Mbps) ATM Isochronous networks

Table 1.2   Three generations of multimedia workstations.

isochronous and ATM networks . The authoring tools are based on object-oriented multimedia integrated into operating systems.

A multimedia system consists of three key elements:

- multimedia hardware,

- operating system and graphical user interface, and

■ multimedia software development and delivery tools.

A multimedia workstation and its components are shown in Figure 1.3. The main subsystems that differentiate a multimedia workstation from a traditional (non-multimedia) workstation include: CD-ROM device, video and audio subsystems, and multimedia related hardware (including image, audio, and video capture, storage, and output equipment).

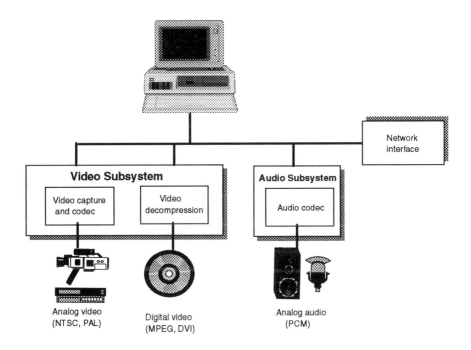

Figure 1.3 Components of a multimedia workstation.

CD-ROM

Besides standard computer peripherals, a multimedia workstation contains a CD-ROM drive, a digital storage medium used for multimedia software distribution. Due to their large sizes, preauthored multimedia files (such as video clips, games, encyclopedias, photo CDs, and others) are stored and distributed on CD-ROMs. A CD-ROM is a read-only, digital medium, whose mastering is expensive, but whose replication is relatively cheap. Its current capacity is about 600 MB, its access times are less than 400 msec, and its transfer rate is 300 KB/sec.

Various CD standards have recently been developed. We will discuss next the most important ones. A detailed overview of CD standards can be found in [KP94]. The relationship among various standards is shown in Figure 1.4.

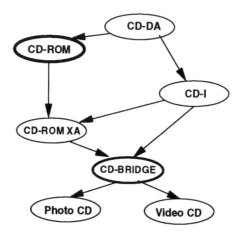

Figure 1.4   The relationship between various CD standards.

CD-I (CD-Interactive) is one of the first CD-ROM products, whose speci-fication defines formats for the representation of audio/visual material. The standard comprises several quality levels of audio coded using the ADPCM (Adaptive Differential Pulse Code Modulation) technique, and various repre-sentations of still images, animations, and motion video. The CD-I disc may also contain CD-DA audio, described next.

CD-DA (CD-Digital Audio) is the functional specification for digital audio. It is based on sound sampled at a frequency of 44.1 KHz and linearly quantized using the PCM (Pulse Code Modulation) technique at 16 bits/sample for each of the left and right audio channels. This produces a data stream of 1.4112 Mbits/sec.

CD-ROM XA (CD-ROM Extended Architecture) represents an extension of the CD-ROM specification with CD-I compatible audio and video formats. The architecture provides a format to interleave audio and image data on the same track. In this way, it provides multimedia applications with synchronized audio and video. It allows CD-ROM XA discs to be played in various operating environments.

CD-MO (CD-Magneto Optic) and *CD-WO* (CD-Write Once) allow recording of information on a special CD using a dedicated recorder. These formats are typically used for low volume production. A CD-MO disc uses a magneto-optic layer for recording information. In contrast, each part of the spiral in CD-WO (sometimes called CD-R which stands for CD Recordable) can be used only once for recording.

CD-Bridge standard is intended to bridge the gap between CD-ROM XA and CD-I. Examples of standards based on CD-Bridge are Photo CD and Video CD .

Photo CD is a standard for storing high-resolution photographic images which can be either as a premastered disc or a CD-WO disc. In the latter case the images can be added to it.

Video CD is another CD Bridge-based standard intended for storing digital video. The video is encoded using MPEG-1 compression standard. A video CD disc can contain up to 80 minutes of full-motion video together with the associated audio.

Video Subsystem

A video subsystem is comprised of a video codec which provides compression and decompression of image and video data. It also performs video capture of TV-type signals (NTSC, PAL), from camera, VCR, and laserdisc , as well as playback of full-motion video. The playback part of the system includes logic to decode the compressed video stream and place the results in the display buffer.

An advanced video subsystem may include additional components for image and video processing. For example, the system can contain output connectors for attachment to a monitor, which will allow the user to view live or captured images during the capturing process. An example of such a system is IBM's Video Capture Adapter/A [IBM92]. In addition, sophisticated image processing techniques can be implemented as the part of the video subsystem, which will allow the user to alter the characteristics of the digitized image.

An additional function of the video subsystem may be the mixing of real-time video images with VGA computer graphics originating from the computer system. An example of a system with such function is IBM's M-Motion Video Adapter/A [IBM92].

Audio Subsystem

An audio subsystem provides recording, music synthesis and playback of audio data. Audio data is typically presented in one of three forms: (a) analog waveform, (b) digital waveform, and (c) MIDI.

Analog waveform audio is represented by an analog electrical signal whose amplitude specifies the loudness of the sound. This form is used in microphones, cassette tapes, records, audio amplifiers, and speakers.

Digital waveform audio is represented using digital data. The digital audio has significant advantages over analog audio, such as less sensitivity to noise and distortion. However, it involves larger processing and storage capacities. Digital devices which use digital waveform audio format are compact disc, the digital audio tape (DAT), and the digital compact disc (DCD) .

MIDI (Musical Instrument Digital Interface) refers to digital encoding of musical information, where the sound data is not stored, and only the commands that describe how the music should be played are generated.

MIDI gives the highest data compression, is easy for editing, and is widely accepted as a musical data standard. However, it requires a music synthesizer to generate music, as shown in Figure 1.5.

An example of an audio subsystem is IBM's M-Audio Capture and Playback Adapter, shown in Figure 1.6 [IBM92]. The system performs digitization of external audio signals through an A/D converter, and generation of audio signals through a D/A converter. The digital signal processor (DSP) performs audio data compression and some additional audio processing functions, such as mixing and volume control.

Some advanced multimedia systems combine both video and audio subsystems into a unified audio/video subsystem. An example of such a system is IBM's

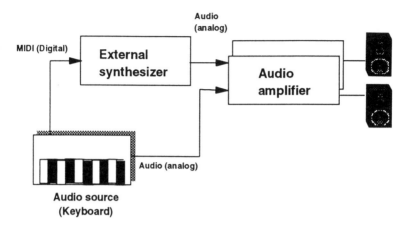

Figure 1.5 MIDI musical data standard offers the highest compression, but requires a synthesizer to generate music.

Action Media II Display and Capture Adapter, which provides the display, capture, and digitized storage of motion video, audio, and still images.

Multimedia Related Hardware

Multimedia related hardware includes video and audio equipment required at multimedia production and/or presentation stages. This equipment can be divided into:

- Image and video capture equipment: still and video cameras, scanners, and video recorders,

- Image and video storage equipment: laserdiscs, videotapes, and optical disks,

- Image and video output equipment: displays, interactive displays, TV projectors, and printers,

- Audio equipment: microphones, audio tape recorders, videotape recorders, audio mixers, headphones, and speakers.

Some of this equipment is described next.

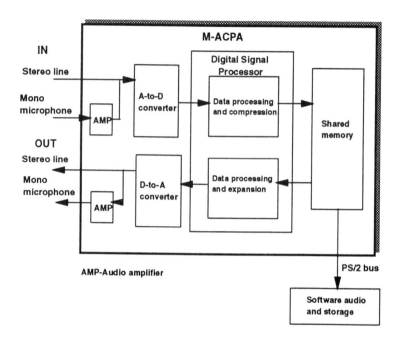

Figure 1.6 The functional diagram of IBM's audio subsystem - M-Audio Capture and Playback Adapter [IBM92].

Video cameras may vary from a fixed TV studio type cameras with facilities for speed matching, to eliminate flicker when capturing from a live TV or computer output, to hand held camcorders that allow complete portability.

The normal consumer cameras use a single image sensor chip, whose output has to be multiplexed to produce three colors: red, green and blue. A three-sensor camera has a separate chip for each color, and each chip produces its own signals, one for red, one for green, and one for blue. Thus, the original RGB signals are not mixed and, if a picture is derived from these signals, it will be of higher quality than one of the single chip camera. The operation of a three-sensor RGB camera is described in Section 4.3.

When a three-sensor camera is used, there are various stages of generation of the video and audio signals, as illustrated in Figure 1.7.

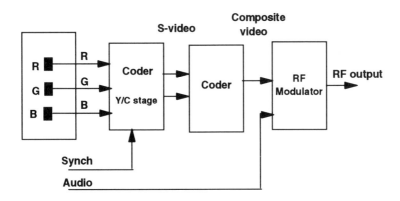

Figure 1.7 Signal stages in a three-sensor camera.

The RGB output produces the highest quality signal, which is characterized by more than 400 lines per frame. The video signal is intercepted at its component level, before red, green and blue signals have been mixed. Besides the RGB output, there are three additional signal outputs: S-video , composite video, and RF output.

The S-video is the next highest quality signal at the output of Y/C stage. At this stage the color is presented using luminance and chrominance components. The resolution at this level is about 400 lines per frame.

Composite video is the simplest form of acceptable video for multimedia applications. A single signal contains the complete picture information, while the audio at that point is represented by a separate signal. The resolution is about 200 lines per frame.

The single RF line output combines the video and audio signals. It provides the lowest quality of video, which is not acceptable for multimedia applications. It represents an RF modulated signal, which is suitable for input into the antenna socket on a conventional TV.

Video recorders are used to store captured video, and they can be either Video Tape Recorders (VTR) or Video Cassette Recorders (VCR). Several formats for storing the video information from the camera to the tape coexist today;

these formats are BETA, VHS, and VIDEO 8. The differences among these formats are in the tape width, signal separation, and tape speed.

In addition, there are several home TV distribution standards worldwide, including PAL, SECAM, and NTSC.

PAL is a European standard which uses a TV scan rate of 25 frames (50 half-frames) per second, and a frequency of 625 lines/frame. The PAL standard is also used in Australia and South America.

SECAM is the French standard, similar to PAL , but it uses different internal video and audio frequencies. Besides France, SECAM is also used in Eastern Europe. Because of the similarity of these two standards, a simple converter can convert videos and audios from SECAM to PAL and vice versa.

NTSC is the USA standard, which is very different from PAL and SECAM standards. The frame rate in NTSC is 30 frames (60 half-frames) per second, and the frequency is 525 lines per frame. The NTSC standard is also used in Canada, Japan, and Korea.

Laserdisc (or videodisc) is a medium for storing analog video signals which are in NTSC , PAL , or SECAM formats. An NTSC system requires 30 frames/second, which gives 30 minutes from a standard ten-inch disc (including a stereo sound track). The PAL/SECAM system requires 25 frames/second, which allows 36 minutes to be stored.

The advantage of videodisc technology over VTR/VCR technology is that each individual frame on the videodisc is addressable. This allows a computer program to access any frame from the disc.

Optical discs are storage media based on optical technology, which can be: Write Once Read Many (WORM) or Read/Write (R/W) optical discs .

The WORM optical disk has been available for years and allows storing the information on the disc only once. R/W (or rewritable) optical disk use the magneto-optical (M-O) technology to provide unlimited read/write capability. They provide large capacity up to 256 MB, and are removable.

Image output equipment includes displays, interactive displays, TV projectors, and printers.

Interactive displays (touch screen) use strain gauge technology, and have additional capabilities to distinguish different levels of pressure. Besides allowing the menu structure to be selected from the screen, they also provide a means to accelerate or slow down some activities by using different pressure on the screen.

TV projectors use the same principles as traditional TV tubes. The video signal is converted into three beams: red, green, and blue, and then projected on a screen. The difference is that in a conventional TV tube the beams are from an electronic gun aimed at a phosphorescent screen, and in the projector TV the beams are of colored light from a very strong light source. Depending where the light source is located, projector TVs can be rear projectors or front projectors.

Multimedia Operating System

A multimedia operating system (MMOS) should provide for sharing of multimedia resources by concurrently running applications, as well as data streaming and synchronization. It should also allow playback and recording of audio, images, and video. A MMOS should also provide software tools for application developers to create standard media control for audio, MIDI, CD-ROM, display, and other peripherals. A generic architecture of a multimedia operating system is shown in Figure 1.8.

Commercial approaches in designing multimedia extensions to operating systems include IBM's Multimedia Presentation Manager/2 (described next), Microsoft's Windows Multimedia Extensions, Apple's QuickTime and others.

IBM's Multimedia Presentation Manager/2 (MMPM/2). The Multimedia Presentation Manager/2 represents extensions to the OS/2 operating system to assist in the development of multimedia applications. The MMPM/2 provides the following main functions:

- Media control interface support. This function provides opening and closing devices, and play, record, stop, pause, and seek functions.

- Record and play audio data, such as CD digital support, waveform audio support, and MIDI synthesizer support.

- Multimedia I/O manager, which allows application independence from data-object format.

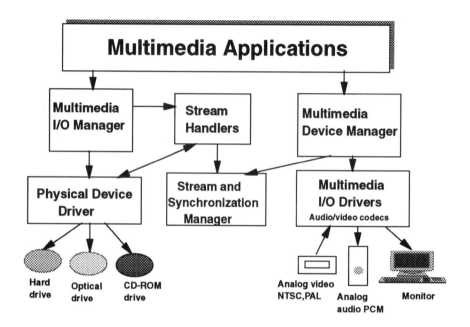

Figure 1.8   A generic architecture of a multimedia operating system.

- Device drivers for audio and video capture and playback adapters.

- Synchronization and streaming functions which provide the application developers with control for synchronizing multimedia data streams, such as audio and video.

- Support for various multimedia data formats, such as AVC audio files and images, MIDI format, PCM and ADPCM audio files, and M-Motion still images.

The architecture of the MMPM/2 is shown in Figure 1.9 [MPM92].

Two main subsystems of the MMPM/2 are the Media Device Manager (MDM) and the Multimedia I/O Manager (MMIO). Applications communicate with hardware devices via MDM and MMIO. The commands are passed from application either through the MDM to a multimedia device driver, and then to the physical device driver (PDD), or through the MMIO manager, a stream

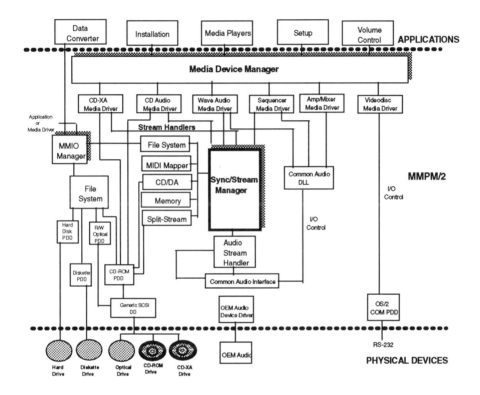

Figure 1.9 The architecture of the IBM's Multimedia Presentation Manager/2.

handler and then to the PDD. Multimedia data is treated as a stream and is controlled through a set of stream handlers and the Sync/Stream manager.

The MMPM/2 supports video from a number of sources, including animation, digital video, videotape, and videodisc. The MMPM/2 implements the concept of software connectors to connect data sources and data targets. There are 17 types of connectors, including MIDI stream, CD stream, headphones, speakers, microphone, video in, video out, and others, that are used to access device-dependent functions. These connections are established automatically when a multimedia device driver is opened.

The sync/stream subsystem is comprised of the sync/stream manager and several stream handlers, including dynamic link libraries and device drivers that control the behavior of individual data streams. The sync/stream subsystem provides a set of functions that control real-time event synchronization and continuous, real-time data streaming. The data streaming function allows multimedia applications to transport large amounts of data continuously between two devices (for example, PCM waveform data from a hard disk to an audio capture card for output), or between memory and a device. The multimedia application creates and starts a stream, and then the sync/stream subsystem takes the responsibility to transport the data between devices.

The sync/stream manager provides stream resource management by connecting a stream source to an appropriate stream target. Stream handlers are responsible for controlling the flow of application data in a continuous, real-time manner. Each handler can establish multiple data stream instances, where each stream involves data of specific type: for example MIDI or ADPCM .

The sync/stream subsystem also provides synchronization of multiple events. For example, in a multimedia application two events, such as displaying a bit-mapped image and playing an audio waveform, should begin at the same time. From a standard OS/2 application perspective, it may appear that the problem can be solved by simply creating two independent threads, which will control these two events, and dispatch them at the same time. However, this approach will not guarantee that both events will occur within a specified time deadline.

The sync/stream subsystem provides synchronization of multiple events including digital video, digitized image, MIDI audio, as well as user-oriented and system-driven events. The solution is based on master/slave relationship between events to be synchronized. There is one object, the synchronization master, which controls the behaviors of one or more data streams, which are slaves. The synchronization master transmits real-time information to all the slaves through the sync/stream manager, giving the slaves the current time in 1/3 msec time units. This time information, called the sync pulse, allows each slave stream handler to adjust the activity of that stream so that synchronization can be maintained.

Figure 1.10 illustrates synchronization between two streams: (1) the file system stream handler is streaming waveform audio data from a CD-ROM device to the waveform stream handler, and (2) the memory stream handler is acting at the source of a MIDI sequence stream, whose target is the audio stream handler, attached to an audio adapter [MPM92].

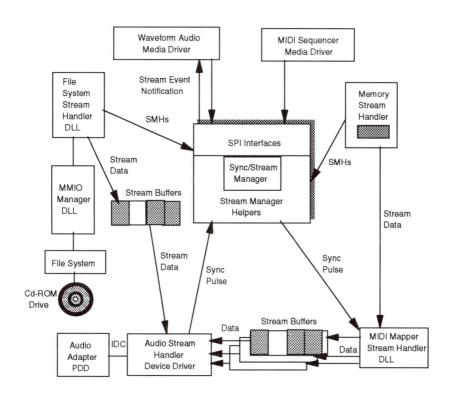

Figure 1.10 Synchronization of two streams in the MMPM/2 system.

Synchronization is done by the audio stream handler, using interrupts it receives from the audio adapter physical device driver (PDD). The waveform stream is the synchronization master, and the MIDI sequence stream is the synchronization slave. Sync pulses are triggered by the audio adapter hardware interrupts, and then passed along to the audio stream handler device driver. The sync/stream manager propagates it to the slave, the audio stream handler, which periodically adjusts MIDI playback tempo to maintain synchronization with the waveform master stream.

A detailed synchronization algorithm, applied in the MMPM/2 system, is described in Section 2.2.4.

1.3   DISTRIBUTED MULTIMEDIA SYSTEM ARCHITECTURES

The main components of a distributed multimedia system, shown in Figure 1.11, are:

- Information (content) providers,

- A wide area network, and

- Multimedia clients.

Various issues in such a distributed multimedia system have been addressed, such as media synchronization, multimedia data storage, indexing , and retrieval , video compression, and networking requirements for transmitting continuous media [Fur94, FM95].

Information servers are connected to the multimedia users through a network consisting of switches and a transmission medium. The transmission medium may be a coaxial cable or a fiber optic channel. Cable companies already have a large network of coaxial cable, albeit for one-way delivery of video. Telephone companies have been upgrading their long haul networks with fiber optics.

Today, cable and telephone systems use different topologies and technologies to deliver their services. The phone system is switched, symmetrical, and interactive. The backbone (trunk link) is typically fiber carrying digital signals, while twisted-pair copper wires carrying analog signals are used to deliver services into homes and businesses. The cable system is unswitched and distributive, built on a backbone of analog fiber and satellites. It uses coaxial cables to connect to customer sites [Rei94]. Both systems are shown in Figure 1.12.

In contrast to telephone and cable systems, the Internet is another network infrastructure, which is used as a government-subsidized experimental electronic communication network. The is still hard to use, does not support billing or transmission of real-time data, and can be both expensive and difficult to access [Rei94, Pre93]. The Internet is still one of the candidates for the data superhighway, due to a growing number of users. However, a future unification of cable and phone systems could reduce the importance of the Internet. In the future, local architectures of both cable and phone systems may become almost identical. Both systems will perhaps become switched and symmetrical. A hybrid fiber/coax network will transmit two-way voice, data, and cable TV

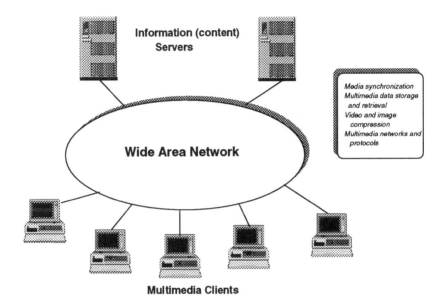

Figure 1.11 A general architecture of a distributed multimedia system.

services. Backbone networks, such as the public switched telephone network or private computer networks, will be connected at the central phone office or cable TV headend.

Table 1.3 compares features of present phone, cable, and Internet networks, critical for the deployment of distributed multimedia applications, such as video-on-demand and interactive television [F+95].

At the present time, telephone, cable and the Internet systems use different communication architectures and standards, as illustrated in Table 1.4. This table also presents expected communication topology and protocols of future unified cable/phone systems.

A detailed discussion on network requirements to support multimedia data transfer is presented in Section 2.1.

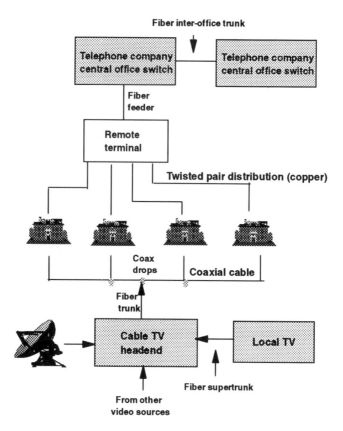

Figure 1.12 Present telephone and cable network architectures use different topologies and technologies.

1.3.1 Multimedia Server Architectures

Sharing multimedia content in a distributed computing environment poses unique challenges to the capabilities of traditional servers. New multimedia server solutions are now emerging that optimize the server's support for sharing multimedia data. To facilitate shared access to multimedia data, multimedia servers need to control very large volumes of continuous data streams, the media devices that store these data streams, and the networks and communications that distribute these data streams to and from the users in real time.

	TELEPHONE	CABLE	INTERNET	Importance for ITV systems
Bandwidth	**Very low**	**High**	**Very low**	**Very important**
Affordable	**High**	**Medium**	**Low**	**Very important**
Easy of use	**High**	**High**	**Low**	**Very important**
Billing	**High**	**Medium**	**Low**	**Very important**
Availability	**High**	**Medium**	**Medium**	**Important**
Information content	**Very low**	**Low**	**High**	**Important**
Security	**High**	**Very low**	**Low**	**Important**
Openness	**Medium**	**Low**	**High**	**Less important**

Table 1.3   Features of three candidates for the data superhighway.

Multimedia data can be represented and stored in either analog or digital form. Analog servers control the devices that store analog data, such as laserdiscs and TV tuners, while digital servers control the digital data by retrieving it from the files or databases. IBM's Ultimedia Video Delivery System (VDS/400), developed for the AS/400 system, is an example of an analog multimedia server [MDC92].

Digital multimedia servers offer a number of advantages including the ability to incrementally add, delete, or edit stored multimedia content, through the use of digital video editing techniques. Digital servers also allow multiple clients to have concurrent access to the same media devices.

A basic architecture of a digital multimedia server at the information provider level consists of a central processing unit, a large capacity store, and a network adapter to a wide-area network (such as ATM) , as illustrated in Figure 1.13.

	TELEPHONE (Today)	CABLE (Today)	INTERNET	CABLE/TELCO (Future)
MEDIA/ BACKBONE	Digital fiber optic (97%)	Satellite, analog fiber optic	NSFnet (T3), other telcos	Analog/digital fiber optic, satellite
MEDIA/ LOCAL	Copper wire wireless	Coaxial cable	Copper wire, Switched-56, T1/FT1	Coaxial cable, fiber optic copper wire, two-way radio
TOPOLOGY	Circuit-switched star	Unswitched, trunk and branch	Packet-switched, routed	Switched/ unswitched, star
PROTOCOLS	POTS, ISDN, ATM	Proprietary analog	TCP/IP	Analog, ADSL, ATM
KEY USERS	Everybody	60% of U.S. households	Government, academia, business	Everybody

Table 1.4   Topologies and protocols of candidate networks for the data superhighway.

The disk storage system is typically based on RAID technology (Redundant Arrays of Inexpensive Discs). Video programs are stored in MPEG compressed form, requiring 1-2 Mb/s transmission for MPEG-1 and 2-40 Mb/s for MPEG-2 videos. The data is retrieved from the disk and transmitted through the network at rates exceeding the real-time rates, which allows simultaneous transmission of a group of video segments. For example, a 60 second MPEG-1 compressed video segment requires 90 Mbits of storage (1.5 Mb/s x 60 s = 90 Mb). The segment can be delivered in an interval of 0.6 seconds at the 150 Mb/s transfer rate (150 Mbps/90 Mb = 0.6 s), which is a hundred times faster than the real-time rate.

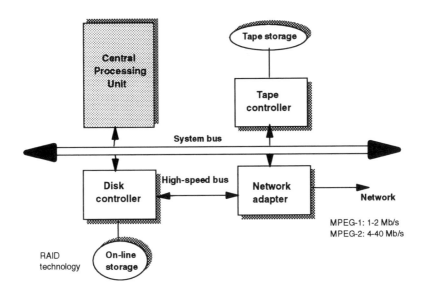

Figure 1.13   The architecture of a digital multimedia server.

However, with MPEG-2 technology, providing HDTV-quality of video, the requirements for the multimedia server become stringent. In addition, large super servers should be able to serve thousands of viewers simultaneously, and therefore they must have large switching capacity to instantly connect data from disks to any output channel. Therefore, large super servers will be built around powerful massively parallel computers, which will provide high I/O throughput.

nCUBE-Based Oracle Media Server

An example of a super multimedia server is a system based on the nCUBE massively parallel computer. The system runs Oracle Media Server software and it is estimated that it will be capable of providing 20,000 simultaneous video streams, thus serving about 60,000 households [Buc94]. The nCUBE parallel computer is scalable from 8 processors and 3 GB of storage capacity to 8192 nodes and 14 TB of disk capacity. Each node can accommodate up to 64 MB of main memory, so the system can cache more than 100 feature-length movies in the processor's memory, providing instantaneous access to the most popular titles. Less frequently viewed titles are striped across multiple disks and retrieved in parallel. The system has up to 1024 I/O channels and supports more than 7000 disk drives in parallel.

The Oracle Media Server software is designed to exploit the high I/O capacity and hypercube interprocessor communication topology of the nCUBE computer, which acts as a large switching system. Each node retrieves a portion of a movie from disk or cache and dispatches digital video streams through the system's common network interfaces to the network [Buc94].

Silicon Graphics Media Server

Silicon Graphics (SGI) is taking a different approach by designing a cluster of 14 SGI servers on an FDDI ring [Pre93]. Two of the machines in the cluster function primarily as session managers and transactions processors, and the others deliver MPEG-compressed video. The video servers have 1.2 Gb/s backplanes, and can have up to 24 processors, 12 GB memory, and several hundred gigabytes of disk. The FDDI ring is used for control and coordination within the cluster. It is estimated that the system will be capable of providing 50 popular movies on-line, with the top 10 on multiple servers. Another 5,000-10,000 movies will be available on tapes. Small video segments will be transferred from the video servers to an AT&T ATM switch. The network will terminate with set top boxes designed by SGI and Scientific Atlanta.

Hewlett-Packard Media Server

Hewlett-Packard (HP) has developed its video server from the ground up rather than retrofitting a computer or file server [FKK95]. The HP video server is optimized for I/O performance. The heart of the server is a Video Transfer Engine (VTE), which has a modular and scalable architecture. It can be scaled in the number of simultaneous video streams supported, and in on-line and off-line storage capacity. The server supports hierarchical management for tapes and magnetic discs, so that along with frequently watched videos, less frequently watched videos can also be supported at a reduced cost. The HP video server can support 75 to 10,000 simultaneous MPEG-1 and MPEG-2 streams at 1 to 6 Mb/s. The target average latency is less than 300 msec for standard services, and less than 100 msec for special gaming services.

Other examples of commercially-developed video servers include Philips/BTS, DEC, Data General, EMC, IBM, Microsoft, Starlight Networks, and Sun.

1.4   ABOUT THE BOOK

In this book we describe critical topics in multimedia systems - image and video processing techniques and their implementations. These techniques include:

- Image and video compression techniques and standards, and

- Image and video indexing and retrieval techniques.

Part II of the book, consisting of 7 chapters, covers image and video compression techniques and standards, their implementations and applications. As mentioned earlier in this chapter, multimedia data (specifically video and images) require efficient compression techniques in order to be stored and delivered in real-time. Video and image compression is therefore a crucial element of an effective multimedia system.

Most digital images contain a high degree of redundancy, which means that an efficient compression technique can significantly reduce the amount of information needed to store or transmit them. This redundancy can be found between single pixels, between lines, or between frames, when a scene is stationary or slightly moving.

Considerable research has been devoted to compression, and compression techniques have been developed. In this book we will concentrate on the most popular techniques which have recently become standards for multimedia compression. These techniques and standards, shown in Figure 1.14, are:

- JPEG - for full color image compression,

- Px64 or H.261 - for video-based communications, and

- MPEG - for intensive applications of full-motion video.

These three compression techniques are described in Chapters 5, 6, and 7. Experimental data are also provided.

Some other promising multimedia compression techniques, such as fractal-based compression, subband coding, wavelet-based compression, and DVI compression, are briefly introduced in Chapter 8.

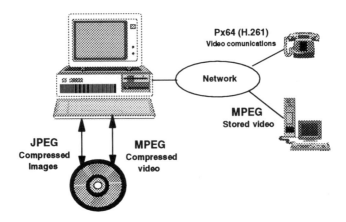

Figure 1.14   Compression standards and their typical applications.

Another aspect of compression deals with the implementation of codecs (compression/decompression devices), which is discussed in Chapter 9. The implementations range from pure software solutions to hybrid and a variety of hardware solutions using specialized VLSI codec chips. A few typical multimedia applications of compression systems are introduced in Chapter 10.

In Part III attention shifts to the *semantic* nature of image and video source material, and how that material may be effectively indexed and retrieved. Chapter 11 begins by concentrating on static images. It discusses how the expressiveness of queries may be expanded beyond what may be represented as text-based propositions to accommodate features of those images and how knowledge of those features may facilitate the indexing of images in a database. Chapter 12 moves beyond images to full-motion video, again addressing questions of just what structural features are and how they contribute to effective indexing. Because so many video data sources are now likely to be stored and maintained in a compressed representation, Chapter 13 demonstrates how those representations can often facilitate structural analysis. Compression is, after all, a form of signal processing. If its transforms yield feature information which is particularly relevant to structural analysis, then that information should be exploited, rather than "reinvented" from the original raster data. Finally, Chapter 14 pulls together most of the key principles of Part III with an extended discussion of a case study — the structural analysis of broadcast television news.

Figure 1.15 provides a graphic summary of the general approach to video indexing and retrieval upon which Part III is based. The emphasis of this approach
is on *content abstraction*, rather than the management of massive quantities of
video resources by a database system.

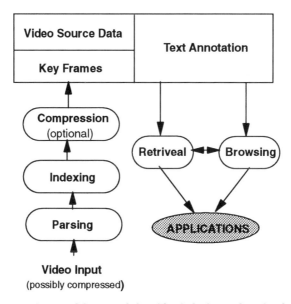

Figure 1.15 A general framework for video indexing and retrieval.

The most effective way to abstract the content of a video for information management tasks will be in the form of static images, such as extracted key frames
which effectively summarize the content of that video. Thus, if a video is to
be stored in a database, it will first be subject to structural analysis (parsing)
to extract those images which will then be indexed on the basis of both their
visual features and text annotations. Both those images and the source video
itself can then be installed in the database, most likely in compressed form.

Recovering information from such a database will, in general, involve a tightly
coupled integration of retrieval and browsing . The purpose of retrieval is
to compose and satisfy precisely formulated questions, but browsing is often
necessary first before one begins to understand just how a question should be
formulated. Similarly, very large databases may provide very large responses
to queries; so browsing is likely to be just as important after a query has been
processed as before it has been articulated.

The reader should be cautioned, however, that all of this is very much work in progress. Very few video databases have been implemented, and most of them deal with far fewer items than are handled by many massive text-based databases. Once we actually start to work with such databases, our views on what they can do are bound to change. Nevertheless, we hope that most of the structural principles which are discussed in Part III are general enough to sustain the coming tides of uncertainty in the sea of multimedia computing.

2

MULTIMEDIA NETWORKING AND SYNCHRONIZATION

In a typical distributed multimedia application, multimedia data must be compressed, transmitted over the network to its destination, and decompressed and synchronized for playout at the receiving site. In addition, a multimedia information system must allow a user to retrieve, store, and manage a variety of data types including images, audio, and video. In this chapter we present fundamental concepts and techniques in the areas of multimedia networking and synchronization.

2.1   MULTIMEDIA NETWORKING

Many applications, such as video mail, videoconferencing , and collaborative work systems, require networked multimedia. In these applications, the multimedia objects are stored at a server and played back at the clients' sites. Such applications might require broadcasting multimedia data to various remote locations or accessing large depositories of multimedia sources.

Required transmission rates for various types of media (data, text, graphics, images, video, and audio), are shown in Table 2.1 [Roy94].

Traditional LAN (Local Area Network) environments, in which data sources are locally available, cannot support access to remote multimedia data sources for a number of reasons. Table 2.2 contrasts traditional data transfer and multimedia transfer.

INFORMATION TYPE	BIT RATE	QUALITY AND REMARKS
DATA	Wide range of bit rates	Continuous, burst and packet-oriented data
TEXT	Several Kbps	Higher bit rates for downloading of large volumes
GRAPHICS	Relatively low bit rates Higher bit rates -100 Mbps and higher	Depending on transfer time required Exchange of complex 3D computer model
IMAGE	64 Kbps Various Up to 30 Mbps	Group-4 telefax Corresponds to JPEG std High-quality professional images
VIDEO	64 -128 Kbps 384 Kbps - 2 Mbps 1.5 Mbps 5-10 Mbps 34/45 Mbps 50 Mbps or less 100 Mbps or more	Video telephony (H.261) Videoconferencing (H.261) MPEG-1 TV quality (MPEG-2) TV distribution HDTV quality Studio-to-studio HDTV video downloading
AUDIO	nx64 Kbps	3.1 KHz, or 7.5 KHz, or hi-fi baseband signals

Table 2.1   Typical transmission rates of various information types in multi-media communication.

Multimedia networks require a very high transfer rate or bandwidth even when the data is compressed. For example, an MPEG-1 session requires a bandwidth of about 1.5 Mbps. MPEG-2 will take 2 to 10 Mbps, while the projected required bandwidth for HDTV is 5 to 20 Mbps. Besides being high, the transfer rate must also be predictable.

The traffic pattern of multimedia data transfer is stream-oriented, typically highly bursty, and the network load is long and continuous. Figure 2.1 shows the ranges of the maximum bit-rate and utilization of a channel at this rate for some service categories [WK90]. The multimedia networks carry a heterogeneous mix of traffic, which could range from narrowband to broadband, and from continuous to bursty.

CHARACTERISTICS	DATA TRANSFER	MULTIMEDIA TRANSFER
DATA RATE	low	high
TRAFFIC PATTERN	bursty	stream-oriented highly-bursty
RELIABILITY REQUIREMENTS	no loss	some loss
LATENCY REQUIREMENTS	none	low, e.g. 20 msec
MODE OF COMMUNICATION	point-to-point	multipoint
TEMPORAL RELATIONSHIP	none	synchronized

Table 2.2   Traditional communications versus multimedia communications.

Traditional networks are used to provide error-free transmission. However, most multimedia applications can tolerate errors in transmission due to corruption or packet loss without retransmission or correction. In some cases, to meet real-time delivery requirements or to achieve synchronization, some packets are even discarded. As a result, we can apply lightweight transmission protocols to multimedia networks. These protocols cannot accept retransmission, since that might introduce unacceptable delays.

Multimedia networks must provide the low latency required for interactive operation. Since multimedia data must be synchronized when it arrives at the destination site, networks should provide synchronized transmission with low jitter .

In multimedia networks, most communications are multipoint, as opposed to traditional point-to-point communication. For example, conferences involving more than two participants need to distribute information in different media to each participant. Conference networks use multicasting and bridging distribution methods. Multicasting replicates a single input signal and delivers it to multiple destinations. Bridging combines multiple input signals into one or more output signals, which is then delivered to the participants [AE92].

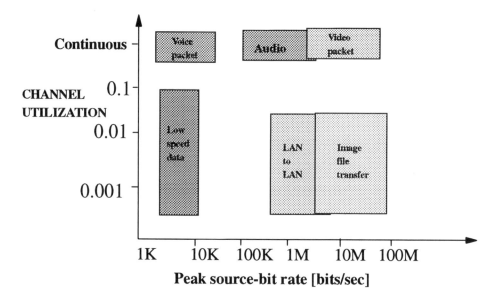

Figure 2.1   The characteristics of multimedia traffic.

2.1.1   Traditional Networks

Traditional networks do not suit multimedia. *Ethernet* provides only 10 Mbps, its access time is not bounded, and its latency and jitter are unpredictable. *Token-ring networks* provide 16 Mbps and are deterministic; from this point of view, they can handle multimedia. However, the predictable worst-case access latency can be very high.

An FDDI network provides 100 Mbps bandwidth, which may be sufficient for multimedia. In the synchronized mode, FDDI has low access latency and low jitter. FDDI also guarantees a bounded access delay and a predictable average bandwidth for synchronous traffic. However, due to the high cost, FDDI networks are used primarily for backbone networks, rather than networks of workstations.

Less expensive alternatives include enhanced traditional networks. Fast Ethernet, for example, provides up to 100 Mbps bandwidth. Priority token ring is another system.

In *priority token ring networks* the multimedia traffic is separated from regular traffic by priority, as shown in Figure 2.2. The bandwidth manager plays a crucial role by tracking sessions, determining ratio priority, and registering multimedia sessions. Priority token ring (PTR) works on existing networks and does not require configuration control [FM95].

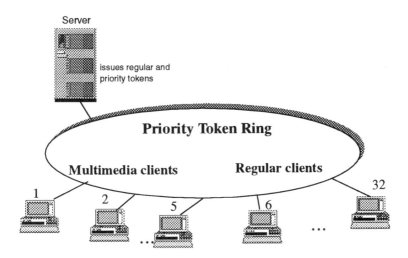

Figure 2.2   Priority token ring can be used in multimedia applications. Five clients with the highest priorities are performing multimedia data transfer, while the remaining 27 clients are performing regular data transfer.

The admission control in PTR guarantees bandwidth to multimedia sessions; however, regular traffic can experience delays. For example, assume a priority token ring network at 16 Mbps that connects 32 nodes. When no priority scheme is set, each node gets an average of 0.5 Mbps of bandwidth. When half the bandwidth (8 Mbps) is dedicated to multimedia, the network can handle about 5 MPEG sessions (at 1.5 Mbps). In that case, the remaining 27 nodes can expect about 8 Mbps divided by 27, or 0.296 Mbps, about half of what they would get without priority enabled.

Crimmins [Cri93] evaluated three priority ring schemes for their applicability to videoconferencing applications: (1) equal priority for video and asynchronous packets, (2) permanent high priority for video packets and permanent low priority for asynchronous packets, and (3) time-adjusted high-priority for video packets (based on their ages) and permanent low priority for asynchronous packets.

The first scheme, which entails direct competition between video conference and asynchronous stations, achieves the lowest network delay for asynchronous traffic. However, it reduces the video conference quality. The second scheme, in which video conference stations have the permanent high priority, produces no degradation in conference quality, but increases the asynchronous network delay. Finally, the time-adjusted priority system provides a trade-off between first two schemes. The quality of videoconferencing is better than in the first scheme, while the asynchronous network delays are shorter than in the second scheme [Cri93].

Present optical network technology can support the Broadband Integrated Services Digital Network (B-ISDN) standard, expected to become the key network for multimedia applications [Cla92]. B-ISDN access can be basic or primary. Basic ISDN access supports 2B + D channels, where the transfer rate of a B channel is 64 Kbps, and that of a D channel is 16 Kbps. Primary ISDN access supports 23B+D channels in the US and 30B+D channels in Europe.

The two B channels of the ISDN basic access provide 2 x 64 Kbps, or 128 Kbps of composite bandwidth. Conferences can use part of this capacity for wideband speech, saving the remainder for purposes such as control, meeting data, and compressed video. Figure 2.3 shows the composition of various channels [Cla92].

2.1.2   Asynchronous Transfer Mode (ATM)

Asynchronous Transfer Mode (ATM) [Bou92, SVH91, KMF94, DCH94] is a packet oriented transport mechanism proposed independently by Bellcore and several telecommunication companies in Europe. ATM is the transfer mode recommended by CCITT for implementing B-ISDN. ATM is considered the network of the future and is meant to support applications with varying data rates.

The ATM network provides the following benefits for multimedia communications:

- It can carry all kinds of traffic, and

- It can operate at very high speeds.

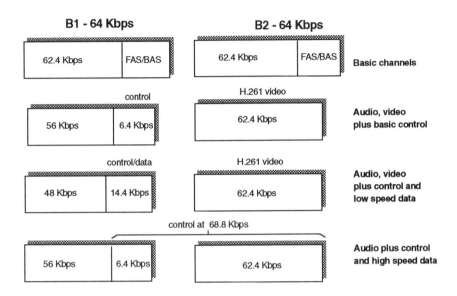

Figure 2.3 Composition of two B channels of the ISDN network for multi-media applications.

The ATM network can carry integrated traffic because it uses small fixed size cells, while traditional networks use variable-length packets, which can be several KB of size.

The ATM network uses a connection-oriented technology, which means that before data traffic can occur between two points, a connection needs to be established between these end points using a signaling protocol. An ATM network architecture is shown in Figure 2.4.

Two major types of interfaces in ATM networks are the User-to-Network Interface (UNI), and the Network-to-Network Interface, or Network-to-Node Interface (NNI).

An ATM network comprises of a set of terminals and a set of intermediate nodes (switches), all linked by a set of point-to-point ATM links, as illustrated in Figure 2.4. The ATM standard defines the protocols needed to connect the terminal and the nodes; however it does not specify how the switches are to be implemented.

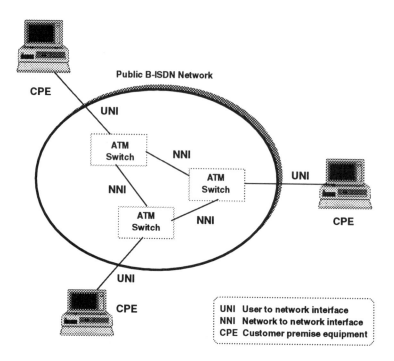

Figure 2.4   The architecture of an ATM network consists of a set of terminal nodes and a set of intermediate nodes (switches).

ATM Cells

The basic transport unit in an ATM network is the cell. Cells are fixed length packets of 53 bytes. Each cell has a 5 byte header and 48 byte payload as shown in Figure 2.5a. The header consists of information necessary for routing. The header doesn't contain a complete destination address. The cells are switched by a switching node using the routing tables which are set up when the network is initiated. The header consists of several fields as shown in Figure 2.5b.

The Generic Flow Control (GFC) field is used for congestion control at the User-to-Network Interface to avoid overloading. The Virtual Path Identifier/Virtual Channel Identifier (VPI/VCI) fields contain the routing information of the cell. The Payload Type (PT) represents the type of information carried by the cell. The CLP field indicates cell loss priority, i.e. if a cell can be dropped or not in case of congestion. The Header Error Control (HEC) field is used

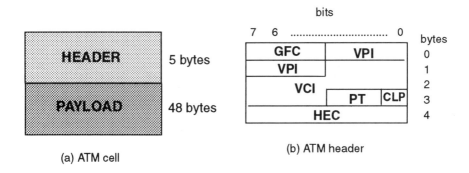

Figure 2.5 Components of (a) ATM cell, (b) ATM header.

to detect and correct the errors in the header. ATM does not have an error correction mechanism for the payload.

ATM Connections

In an ATM network a connection has to be established between two end points before data can be transmitted. An end terminal requests a connection to another end terminal by transmitting a signaling request across the UNI to the network. This request is passed across the network to the destination. If the destination agrees to form a connection, a virtual circuit is set up across the ATM network between these two end points.

These connections are made of *Virtual Channel (VC)* and *Virtual Path (VP)* connections. These connections can be either point-to-point or point-to- multipoint. The basic type of connection in an ATM is VC. It is a logical connection between two switching points. VP is a group of VCs with the same VPI value. The VP and VC switches are shown in Figure 2.6.

Let us consider an example of setting up a VC connection between two nodes N1 and N2, as shown in Figure 2.7. The process of establishing the VC connection consists of the following steps:

- The end terminal N1 sends the request to the UNI.

- The UNI forwards the request to the network.

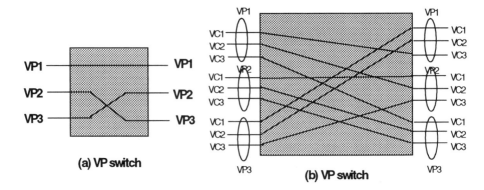

Figure 2.6   Virtual connections in ATM: (a) Virtual Path (VP) switch, and
(b) Virtual Channel (VC) switch.

- Assume that the network selects the following route A-B-C-D. Each of the
  four nodes will use an unused VC value for the connection. For example,
  the intermediate node A chooses VC1, and therefore all the cells from N1
  will have a label VC1 at the output from A.

- The node A sends the cell to the node B. B will change VC1 to VC2 and
  will send VC2 to the node C.

- At the node C, VC2 is associated with VC3 and sent to the node D.

- At the node D, VC3 is associated with VC4. The node D checks if the
  UNI at the terminal node N2 is free. If the UNI is free, the cell with the
  label VC4 is given to N2.

- The terminal node N2 uses now VC4 for its connection to the node D.

- D sends this cell to C, which associates VC4 with VC3, and sends it to B.
  B associates VC3 with VC2 and sends to A. A associates VC2 with VC1
  and delivers the cell to the terminal node N1.

- The connection between terminal nodes N1 and N2 is thus established.

When the transmission of data is completed, N1 sends a message to tear down
the connection, and VCI/VPI values are free to be used by any other connection. Once the connection is established, all the cells travel over the same
virtual channel connection. The cells within a VC are not switched out of

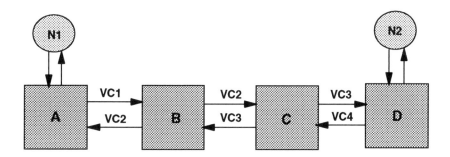

Figure 2.7   ATM connection between two terminal nodes N1 and N2.

order and no retransmissions are done in ATM . This guarantees that all the cells are delivered in sequence to the destination. But, because of buffering at the source, end-to-end delay is not predictable. ATM also does not guarantee that a cell will be delivered to the destination. This is because a cell could be dropped in the case of congestion.

During the call setup phase, a user negotiates network resources such as peak data rate and parameters such as cell loss rate, cell delay, cell delay variation, etc. These parameters are called the Quality of Service (QOS) parameters. Connection is established only if the network guarantees to support the QOS requested by the user. A resource allocation mechanism for establishing connections is shown in Figure 2.8 [SVH91]. Once the QOS parameters for a connection are set up, both the user and the network stick to the agreement.

Even when every user/terminal employs the QOS parameters, congestion may occur. The main cause for congestion is statistical multiplexing of bursty connections. Two modes of traffic operation are defined for ATM : (a) statistical and (b) non-statistical multiplexing [WK90]. In the general ATM node model shown in the Figure 2.9, if the sum of the peak rates of the input links does not exceed the output link rate ($\Sigma P_i \leq L$), the mode of operation is called non-statistical multiplexing; and if ΣP_i exceeds L it is called statistical multiplexing. During connection setup, each connection is allocated an average data rate of that channel instead of the peak data rate. Several such channels are multiplexed hoping that all the channels do not burst at the same time. If several connections of a channel burst simultaneously, congestion might occur.

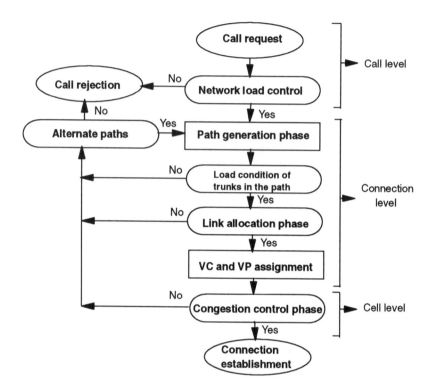

Figure 2.8 A resource allocation mechanism for establishing connections [SVH91].

ATM Protocols

The ATM protocol reference model, shown in Figure 2.10, is similar to the OSI layered model [DCH94]. Communication from higher layers occurs through three layers: the ATM layer, the Physical layer, and the ATM adaptation layer (AAL). The portion of the layered architecture used for end-to-end or user-to-user data transfer is known as the User Plane (U-Plane). Similarly, higher layer protocols are defined accross the ATM layers to support switching - this is referred as the Control Plane (C-Plane). The control of an ATM node is performed by a Management Plane (M-Plane), which is further divided into Layer Management, that manages each of the ATM layers, and Plane Management for the management of all the other planes.

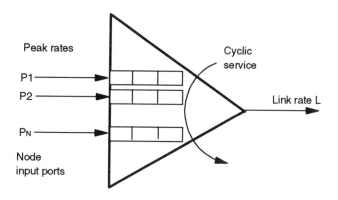

Figure 2.9   General ATM node model.

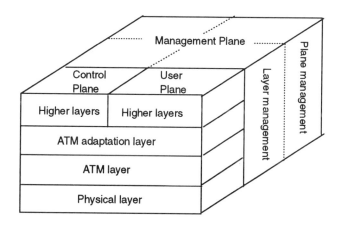

Figure 2.10   The structure of the ATM protocol.

2.1.3   Comparison of Switching Technologies for Multimedia Communications

Besides the ATM technology, other switching technologies considered for multimedia communications include:

■ Synchronous Transfer Mode (STM),

■ Switched Multi-Megabit Data Service (SMDS), called 802.6, and

■ Frame Relay.

STM is a circuit switched network mechanism used to transport packetized voice and data over networks. When a connection is established using STM, the bandwidth allocated for the connection is reserved for the entire duration of the connection even though there may not be data transmission over the entire duration. This is a significant waste of bandwidth which could otherwise have been used by a terminal node waiting for network bandwidth. So this mode limits the maximum number of connections that can be established simultaneously. The ATM network overcomes this limitation by using statistical multiplexing with hardware switching.

SMDS uses a telecommunication network to connect LANs LAN into Metropolitan Area Networks (MANs) and Wide-Area Networks (WANs).

Frame Relay is a network similar to packet switching networks, which consists of multiple virtual circuits on a single access line. It operates at higher data rates than most packet networks, and also gives lower propagation delays.

Table 2.3 compares the properties of these four switching technologies for multimedia communications [RKK94]. The properties compared include support of multimedia traffic, connectivity, performance guarantee, bandwidth, support for media synchronization, end-to-end delay , and congestion control.

According to Table 2.3, it is clear that ATM is the superior switching technology for multimedia communications. ATM is capable of supporting multimedia traffic, it provides various types of connectivity, it guarantees performance, its bandwidth is very large (up to several Gbits/s), its end-to-end delay is relatively low, and it guarantees media synchronization.

2.2 MULTIMEDIA SYNCHRONIZATION

Multimedia systems include multiple sources of various media either spatially or temporally to create composite multimedia documents. Spatial composition links various multimedia objects into a single entity (Figure 2.11a), dealing

Various switching technologies

Parameters		ATM	STM	SMDS (802.6)	Frame Relay
Support of multimedia traffic transfer	Data	Yes	Yes	Yes	Yes
	Audio	Yes	Yes		
	Video	Yes	Yes		
Connectivity	One-to-many	Yes	Limited	Yes	Yes
	Many-to-one	Yes	Limited	Yes	Yes
	Many-to-one	Yes		Yes	Yes
Performance guarantee	Delay for audio/video	Yes	Yes		
	Packet loss for data	Yes	Not applicable	Yes	Yes
Bandwidth		1.5 Mb/s to multi Gb/s	Few bits/s to multi Gb/s	1.5 Mb/s to 45 Mb/s	56 Kb/s to 1.5 Mb/s
Guarantee of media synchronization		Yes	Yes		
End-to-end delay		Low	Lower than ATM	Higher than ATM	Higher than ATM
Congestion control		Required	Not required	Required	Required

Table 2.3   Comparison of switching technologies for multimedia communications. Adapted from [RKK94].

with object size, rotation, and placement within the entity. Temporal composition creates a multimedia presentation by arranging the multimedia objects according to temporal relationship (Figure 2.11b) [LG90].

We can divide temporal composition, or synchronization, into continuous and point synchronization . *Continuous synchronization* requires constant synchronization of lengthy events. An example of continuous synchronization is video telephony, where audio and video signals are created at a remote site, transmitted over the network, then synchronized continuously at the receiver site for playback. In *point synchronization*, a single point of one media block

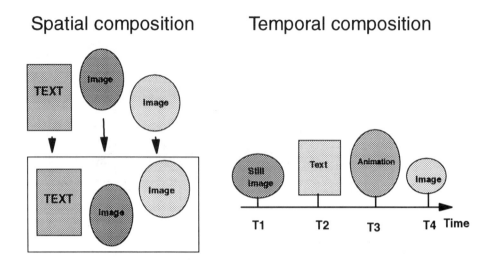

Figure 2.11   Composition of multimedia objects: (a) spatial, (b) temporal.

coincides with a single point of another media block. An example of point synchronization is a slide show with blocks of audio allotted to each slide.

Two further classes of synchronization are *serial* and *parallel* synchronization. Serial synchronization determines the rate at which events must occur within a single data stream (*intramedia synchronization*), while parallel synchronization determines the relative schedule of separate synchronization streams (*intermedia synchronization*).

Figure 2.12 illustrates an example of both serial and parallel synchronization of several streams: audio, video, graphics, text, and animation. Arrows indicate different points of synchronization between different streams. Note that in several cases a single point of two media objects must coincide (point synchronization), while in one case a number of points of two media objects (audio and video) should coincide (continuous synchronization).

2.2.1   Data Location Models

Responsibility for maintaining intermedia synchronization falls onto both the sources and destinations of data, but most techniques rely more on the desti-

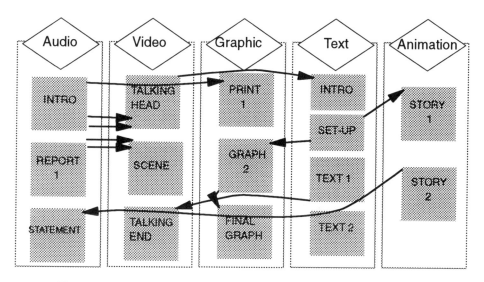

Figure 2.12 Synchronization of several media objects. Arrows indicate points of synchronization.

nations. One topology classification that can determine the required synchronization is based on data location models [LG90, BL91]. Figure 2.13 shows four data location models:

- Local single source. A single source, such as a CD-ROM , distributes the media streams to the playback devices. As long as the devices maintain their playback speed, no synchronization technique is required.

- Local multiple sources. More than one source distributes media streams to the playback devices. An example is a slide show played with music or an audio tape. Synchronization is required within the workstation.

- Distributed single source. One source, such as a videotape, distributes media streams across a network to one or more nodes' playback devices; an example is cable TV. The techniques requires no synchronization other than maintaining the speeds of the playback devices.

- Distributed multiple source. This is the most complex case, where more than one source distributes media streams to multiple playback devices on multiple nodes. This group further breaks down into multiple sources from

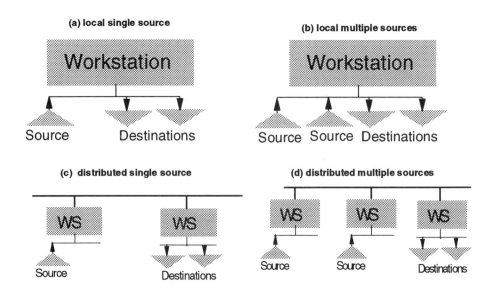

Figure 2.13 Location models for multimedia data: (a) local single source, (b) local multiple sources, (c) distributed single source, (d) distributed multiple sources.

one node distributed to another node (for example, a video call); multiple sources from two or more nodes distributed to another node; multiple sources from one node distributed to two or more nodes (for example, HDTV); and multiple sources from two or more nodes distributed to two or more nodes (for example, a group teleconference).

For the first two location models, local synchronization within the workstation suffices. However, the two cases with distributed sources require more complex synchronization algorithms to eliminate the various causes of asynchrony [LG90]. Figure 2.14 shows an example of video telephony, noting various places within the systems that contribute to asynchrony. The task of synchronization, whether implemented in the network or in the receiver, is to eliminate all the variations and delays incurred during the transmission of multiple media streams and to maintain synchronization among the media streams.

The end-to-end delay of a distributed multimedia system consists of all the delays created at the source site, network, and receiver site. It differs slightly

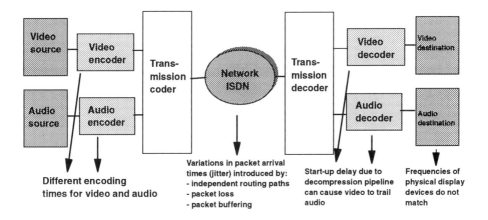

Figure 2.14 Various causes of asynchrony in a video telephone system.

for real-time video and audio and for stored multimedia objects, as shown in Figure 2.15.

FOR CONTINUOUS SYNCHRONIZATION

Tsample   Tencode   Tpacketize   Ttrasmit   Tbuffer   Tdepacketize   Tdecode   Tpresent

FOR STORED OBJECTS (Remote Data Access)

Tquery   Tseek   Taccess   Tpacketize   Ttransmit   Tbuffer   Tdepacketize   Tdecode   Tpresent

Figure 2.15 Definitions of the end-to-end delay: (a) for continuous synchronization, (b) for stored objects.

2.2.2   Quality of Service

Implementing a synchronization algorithm for a specific application requires
specifying the quality of service (QOS) for multimedia communications. The
QOS is a set of parameters that includes speed ratio, utilization, average delay,
jitter, bit error rate, and packet error rate [LG90]. Figure 2.16 illustrates how
to calculate some of these parameters.

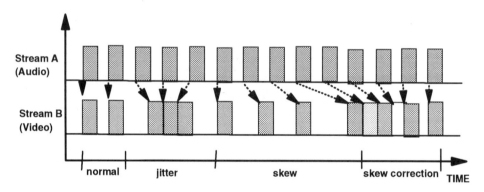

Figure 2.16   Synchronization of two streams showing skew, jitter, utilization
ratio, and speed ratio.

The speed ratio is the quotient of the actual presentation rate divided by the
nominal presentation rate. For the interval (t_0, t_1) in Figure 2.16, the speed
ratio equals 6 over 6, or 1. For the interval (t_1, t_2), the speed ratio equals 4
divided by 6, or 0.67.

The utilization ratio equals the actual presentation rate divided by the available
delivery rate. Ideally, both the speed and utilization ratios equal 1. Duplication
of frames will create a utilization ratio greater than 1, while frame dropping
will cause it to be less than 1.

Jitter is the instantaneous difference between two synchronized streams.

Skew is the average difference in presentation times between two synchronized
objects over n synchronization points. Duplicating a data frame causes the
stream to retard in time (stream lag). Data losses resulting in playout gaps
cause streams to advance in time (stream lead). In Figure 2.16, for the time
interval (t_1, t_2), the skew is 4 divided by 6.

Figure 2.16 also shows how to correct skew by dropping frames. In this example, three shaded frames are dropped to reestablish synchronization.

Two more QOS parameters, *the bit error rate (BER)* and *the packet error rate (PER)*, specify the required reliability of the network. Table 2.4 compares the quality of service required for video telephony and JPEG video transmission.

	Video telephony	JPEG video transmission
Speed ratio	1.0	1.0
Utilization	1.0	1.0
Average delay	0.25 sec	0.2 sec
Minimum jitter	10 msec	5 msec
Maximum BER	0.01	0.1
Maximum PER	0.001	0.01

Table 2.4 Quality of service requirements for video telephony and JPEG video transmission.

2.2.3 Single and Multiple Stream Synchronizations

Synchronization entails evaluating the temporal characteristics of the data streams to be synchronized and correcting all delays and other anomalies [LG91].

To synchronize an event, we first analyze the end-to-end delay, or latency. Then we schedule a retrieval time that allows enough time before the deadline to allow for latency. For example, if the total latency time of retrieving a one-hour video is three minutes, and the customer ordered the video for 7 P.M., set the retrieval time, or the packet production time, for 6:57 P.M. at the latest. Figure 2.17 shows timing of the single event synchronization.

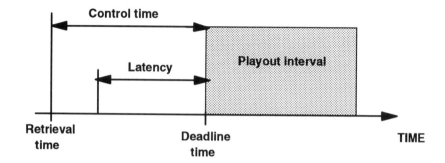

Figure 2.17 Timing for the single-event synchronization.

The client site receives packets of data, created at a source and transmitted over the network, with a random distribution. We can reduce arriving packet delay variance using buffering or some other techniques. Then, synchronization consists of calculating the control time based on multiple deadlines, playout times, and latencies. In this case, time T represents the time required for buffering at the receiving site to smooth out variations in latencies.

Figure 2.18 illustrates the use of the buffering technique in reducing the delay variance of arriving packets. Packets of data, created at a source and transmitted over the network, are received with a random distribution $p(t)$. Using buffering, the reduction of delay variance can be achieved, so the distribution $w(t)$ can be obtained, as shown in Figure 2.18.

We can extend synchronization of single streams of packetized audio and video to the general case of synchronizing multiple streams as well as nonstream data (for example, still images and text) [RG91]. Then, various techniques determine the delays and buffering required to synchronize multiple events for given packet loss probabilities and network delay distributions.

2.2.4   Case Study: Synchronization Algorithm Implemented in MMPM/2

The IBM's Multimedia Presentation Manager/2 (MMPM/2) was introduced in Section 1.2. The Sync/Stream Manager (SSM), shown in Fig. 1.10, is re-

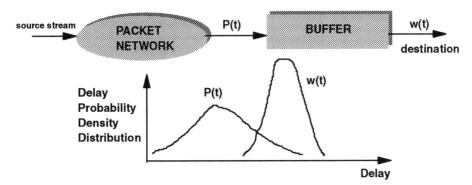

Figure 2.18 The reduction of arriving packet delay variance in single stream synchronization. Buffering is used to create the $w(t)$ probability density distribution of delays from $p(t)$.

sponsible for providing the synchronization of various streams in the MMPM/2 system. It monitors streams to be synchronized (typically audio and video streams), it checks the timing of these sequences making sure that the streams are within the specified tolerance, and, if necessary, adjusts the video output timing in order to correct synchronization.

In the MMPM/2 system (as in most other synchronization solutions) the audio stream is the *master stream* and the video stream is a *slave stream*, which has to be adjusted (synchronized) to the master stream. The adjustment of the video stream is performed by calculating the time when the next video frame has to be displayed. The flow-chart of the MMPM/2 synchronization algorithm, adapted from [Wil94], is given in Figure 2.19.

The algorithm first gets the current time using the system timer (block 1), and, based on the video's frame rate, calculates the time when the next frame should be displayed (block 2). Then, the algorithm determines whether the video is ahead or behind the calculated time (block 3), and it calculates the error in the video stream (block 4). In this way, the algorithm calculates the timing of the video relative to the system timer. If the video stream is within the specified tolerance, the algorithm continues processing the next frame (block 5).

However, if the video stream is ahead or behind the system timer (greater than the tolerance), the algorithm calculates the error between the audio and

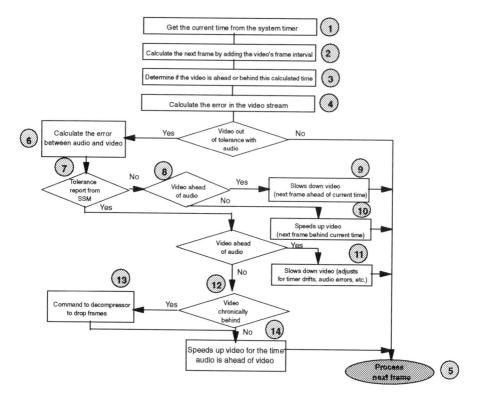

Figure 2.19   The flow-chart of the MMPM/2 synchronization algorithm.

video streams (block 6), and checks the report from the SSM for whether the slave stream (video) is out of tolerance (block 7). If the SSM test is within the tolerance, the algorithm tests whether video or audio is ahead (block 8), and accordingly slows down or speeds up the video stream (blocks 9 and 10, respectively).

However, if the SSM reported an "out of tolerance" error, which means that video and audio are out of synchronization, the algorithm again forces video to get synchronized with audio. In this case, if video is ahead of audio, the algorithm adjusts for timer drifts, audio errors, and other causes (block 11). If audio is ahead of video, the algorithm checks the recent history to determine whether this happened often in the recent past (block 12). If this is the case, which means that video is chronically behind audio, the algorithm sends a command to the decompresser to drop some video frames (block 13). In both cases, the algorithm also adjusts video by speeding up video for the time that audio is ahead of video (block 14).

The frame dropping algorithm, which forces video to speed up in order to catch up with audio, is carefully designed to drop only specific frames in a compressed video. If the MPEG compression algorithm is used, a group of frames (referred as I frames) compress the entire frame, while other groups of frames (referred as P and B frames) are based on predictive coding and save only the portion of a frame that has changed compared to the previous frame. Typically, P and B frames are selected to be dropped. This may cause some visual artifacts; however displaying the I frames will compensate for these artifacts.

3

OVERVIEW OF MULTIMEDIA APPLICATIONS

Multimedia systems suggest a wide variety of applications. Table 3.1 summarizes numerous multimedia applications, including used media and implemented functions [LG90, Fur94]. Two important applications already in use, multimedia conferencing systems and multimedia on-demand services (including interactive television and news on-demand distribution), are briefly described in this chapter. Both applications use multimedia compression techniques, such as MPEG and px64 .

3.1 MULTIMEDIA CONFERENCING

Multimedia conferencing systems enable a number of participants to exchange various multimedia information via voice and data networks. Each participant has a multimedia workstation, linked to the other workstations over high-speed networks. Each participant can send and receive video, audio, and data, and can perform certain collaborative activities. The multimedia conference uses the concept of the *shared virtual workspace*, which describes the part of the display replicated at every workstation.

3.1.1 Components and Functions of a Multimedia Conferencing System

A typical multimedia conferencing (or videoconferencing) system and its components is shown in Figure 3.1 [Mil93]. Videoconferencing stations are equipped with video and audio capture and compress subsystems, and decompress and

APPLICATION	MEDIA	SELECTED FUNCTIONS
Office automation	Images,text,spreadsheet	Composition, filing, comm.
Medical Information Systems	Video (telephony), images	Data acq., comm, filing
Geography	Images, graphics	Data acq., image manip., store
Education/Training	Audio,video,images,text	Browsing,interactivity
Command and Control	Audio (telephony), images	Data acquisition, comm.
Weather	Images,numeric data,text	Data acq., simulation,integration
Banking	Numeric data,text, images	Image archiving
Travel Agents	Audio,video,images,text	Video browsing,communication
Advertising	Video,images	Image composition,enhancem.
Electronic Video Mail	Audio,images,text,video	Communication
Engineering CAD/CAM	Numeric data, text, images	Cooperative work
Consumer Electronic Catalogs	Audio,video,text	Video browsing
Home Video Distribution	Audio,video	Video browsing
Real Estate	Audio,video,images,text	Video browsing, comm.
Digital Libraries	Images,text	Database browsing, query
Legal Information Systems	Image,text	Database query
Tourist Information Systems	Audio,video,text	Video browsing
Newsprint Publication	Image,text	Image,text compression
Dictionaries	Image,text	Database browsing, query
Electronic Collaboration	Audio,video,text	Videoconferencing, comm.
Air Traffic Control	Audio,text,graphics	Concurrency control, comm.
Interactive TV	Video,audio	Interactivity, comm.

Table 3.1   Multimedia Applications and Their Characteristics.

display subsystems. The communication media can be POTS (Plain Old Telephone Systems), LAN or WAN.

The common software architecture of a videoconferencing system is shown in Figure 3.2 [Mil93]. The biggest performance challenge in multimedia conferencing occurs when conference participants continuously transmit video and voice streams. Current research focuses on mixing these streams together to form a composite stream consisting of video and audio streams. Ramanathan

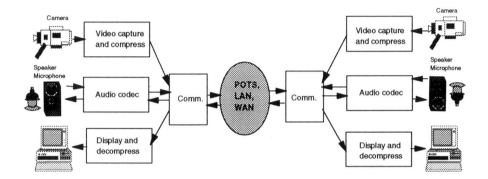

Figure 3.1 Components of a multimedia videoconferencing system.

et al. proposed techniques and protocols for media mixing and an optimal communication architecture for multimedia conferencing [R+91].

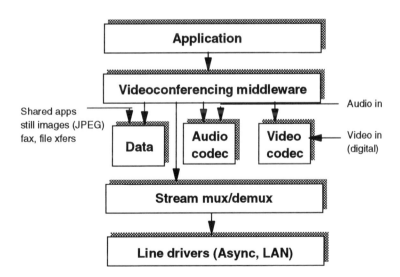

Figure 3.2 Common software architecture of a videoconferencing system.

Multimedia conferencing systems must support a number of functions briefly described next [Gon94].

Multipoint connection set up. The system should be capable to negotiate for network resources and end-to-end conference capabilities.

Dynamic session control. The system should have an ability to add and delete participants to/from an existing conference. Such a change may require modification of the underlying network configuration.

Conference directory. The system should provide a conference directory service that will support conference registration, announcement, and query. The directory should contain various information such as: the title and brief description of the conference, a list of participants, start and end time for the conference, audio and video coding schemes, their protocols and QOS requirements, and shared working space [Gon94].

Automatic conference scheduling and recording. The conference scheduling function combined with resource reservation mechanisms will allow planning of network resources. Automatic conference recording is a useful function which does recoding of conference sessions in a form of multimedia documents.

Conference teardown. The system should be capable to release all reserved resources when the conference is complete.

3.1.2   Network Architectures for Multimedia Conferencing

Several network configurations are proposed and used for multimedia conferencing (Figure 3.3):

- Fully distributed (mesh) network,

- Centralized (star) network,

- Double-star network, and

- Hierarchical network.

The fully distributed network, shown in Figure 3.3a, is based on multiple point-to-point links, and connects each participant with each other. In the multimedia conference each participant sends their media directly to every other

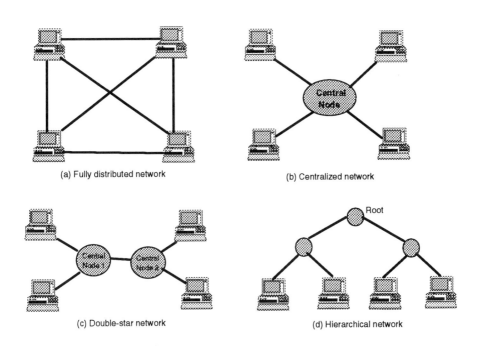

Figure 3.3 Network configurations for multimedia conferencing: (a) Distributed network (mesh), (b) Centralized network (star), (c) Double-star network, (d) Hierarchical network.

participant. The processing and mixing of the media is done at each individual participant's location. This configuration uses the largest network bandwidth and duplicates the work involved in mixing data at every participant's location. On the other hand, this system gives the best (shortest) delay time when considering media transmission and mixing. Its major disadvantage is that with the increased number of conference participants, the point-to-point connections increase rapidly.

The centralized network, shown in Figure 3.3b, consists of a central node connected to every participant of the conference. The central node acts as an intermediate processor and performs media mixing. It receives multimedia data from the participants, mixes or processes the media, and then broadcasts the media back to participants. The centralized system has an advantage that the mixing and processing of media is only done once within the system. The disadvantage of the system is an increased delay time when transmitting composite media from one to another conference participant, since the intermediate processor must wait until all media is received before it begins media mixing and broadcasting.

A double-star network, shown in Figure 3.3c, is an extension of the centralized (star) network. In this system, a central node from one star network is connected to another central node of another star network. A central node is used as a concentrator for several sites communicating via a single bidirectional link with several sites connected to a second central node.

A hierarchical network is another extension of the centralized network. The system consists of a series of intermediate nodes, with one root node and other as internal nodes in a tree structure, as shown in Figure 3.3d. All intermediate nodes are capable of performing mixing and processing, while the leaves in the tree are the conference participants.

Multimedia data is sent up to the tree for mixing and processing. A mixer receives the media from the multiple leaves or mixers below itself, and then transmits the mixed media to the mixer above. The completely mixed data is generated by the root node and then broadcasted directly from the root to the leaves of the tree involved in the conference.

This configuration reduces the network traffic, and therefore the system is capable of handling larger number of conference participants than either centralized or distributed networks.

Multimedia conferencing poses very strict timing requirements. The one-way end-to-end delay is defined as the time from the moment when a conference participant moves or speaks until the motion or sound is perceived by the other participants. This time should be less than 150 milliseconds, which gives the total round-trip delay less than 300 milliseconds for maintaining conversation "face-to-face" [Roy94].

The total one-way "time budget" comprises of four major time components, as illustrated in Figure 3.4 [Mil93]:

- workstation operation,

- workstation transmission time,

- network delay, and

- receiving workstation.

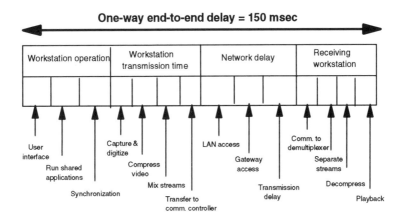

Figure 3.4 The one-way end-to-end delay in videoconferencing.

The delay between audio and video is also very critical and limited to -20 to 40 milliseconds in order to provide good lip-synchronization [Roy94].

Prototype multimedia conferencing systems include MMConf, implemented in the Diamond system [BG91], RTCal (Real Time Calculator), developed at MIT [BG91], Rapport, developed at AT&T [Cla92], the European collaborative projects MIAC (Multipoint Interactive Audiovisual Communications) and MIAS (Multipoint Interactive Audiovisual System) [HKS93], DECspin desktop videoconferencing system [PP93], IBM's Person-to-Person, Intel's Proshare, and others.

3.2 VIDEO-ON-DEMAND APPLICATIONS

Advances in distributed multimedia systems have begun to significantly affect the development of on-demand multimedia services, such as interactive entertainment, video news distribution, video rental services, and digital multimedia libraries [LV94, RVR92]. Various companies realized that fiber-optic networks, coupled with improved computing and compression techniques, would soon be capable of delivering digital movies. Over the past year, a number of alliances have formed between entertainment, cable, phone, and computer companies, with the main focus on video-on-demand (VOD) applications. The information superhighway, digitally compressed audio and video, interactive movies and games, access to databases and services, all on the TV, will change culture and habits of the future generations.

In this section we present two VOD applications: Interactive Television (ITV), and a news on-demand distribution system.

3.2.1   Interactive Television

A well-designed ITV system should be capable of providing the following interactive services [F+95]:

- basic TV,

- subscription TV,

- pay per view,

- video on demand,

- shopping,

- education,

- electronic newspaper,

- digital audio,

- financial transactions, and

- single-user and multi-user games.

Other basic applications to be supported by the ITV include electronic program guide, electronic navigator, and electronic yellow pages.

ITV Network Architectures

Interactive television systems use distributed multimedia architectures introduced in Section 1.3. The main components of an architecture for interactive TV are:

- information (content) servers,

- a network, and

- set top boxes.

Information servers are connected to set top boxes (STBs) at the customer premises through a network consisting of switches and a transmission medium. The transmission medium can be a coaxial cable or a fiber optic channel. Cable companies already have a large network of coaxial cable albeit for one way delivery of video. Telephone companies have been upgrading their long haul networks with fiber optics. However, the telephone cable to the home is still predominately a copper twisted pair. Running fiber to each home is still too expensive, and therefore the most feasible present solution seems to be a Hybrid Fiber Coax (FBC) network, where fiber will be run to a small neighborhood node, and from there a coaxial cable connection will be run to each home.

A general architecture for deployment of an interactive TV system is shown in Figure 3.5. A set of servers offer different types of services. They are connected to the headends of cable network CATV (Community Antenna Television) trees via a wide-area network, such as ATM. In the case of telephone networks, the headend might be a switching office.

The general architecture from Figure 3.5 can be expanded into a hierarchical configuration of multimedia servers and network switches [RR94, F+95]. The system, shown in Figure 3.6, consists of information (service) providers (such as entertainment houses and television stations), who offer various services, network providers who are responsible for media transport over integrated networks, and several levels of storage providers who manage multimedia data storage in multimedia servers and contain network switches.

For illustration purposes, the ITV system from Figure 3.6 will be capable of supporting 100 Metropolitan Area Networks (MANs), each MAN can be linked to an average of 800 headends, and each headend will support about 1000 households. Therefore, the system will support a total 80 million households [NPS94].

Presently, the cable TV system in the U.S. is one-way and based on broadcasting of analog video through a wire. A typical 450 MHz plant uses analog 6 MHz channels, which gives the total capacity of about 70 channels. Their network management and system reliability are relatively primitive. The cable systems tend to be proprietary and not interconnected. Cable companies are currently emphasizing "broadcast" type networks that require minimal switching with

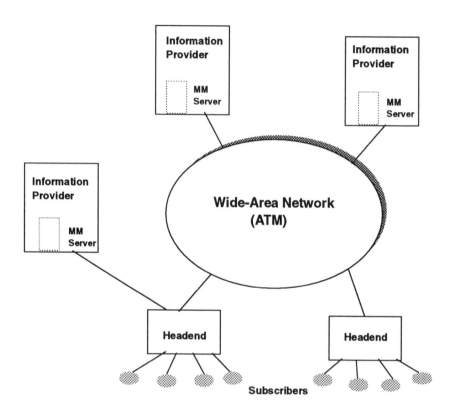

Figure 3.5 A general architecture for interactive TV services consists of information providers, wide-area network, headends, and subscribers.

Near-Video-On-Demand (NVOD) technology. These hybrid digital-analog systems use 750 MHz fiber-to-the node systems that offer a higher number of analog channels and movies in staggered schedules with limited random-access capabilities to users. Cable companies' view in migrating to interactive VOD television is presented in Figure 3.7 [Mil94, F+95].

First, a return path should be incorporated into the present cable system to allow 2-way interactive communications. Then, digitally encoding and compression of video must be provided.

With 64 QAM (Quadrature Amplitude Modulation) it is possible to get 27 Mbps out of a 6 MHz analog channel, while 256 QAM provides a usable bit rate

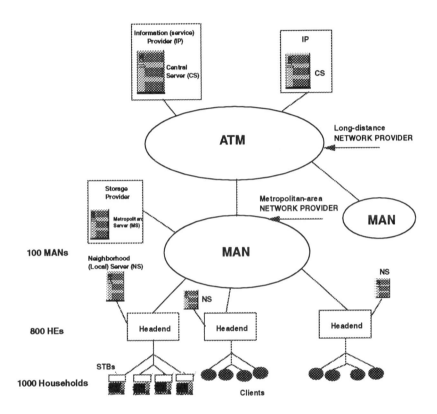

Figure 3.6 Hierarchical configuration of multimedia servers and network switches for interactive television. A typical system will be able to support about 80 millions households in continental U.S: (1 ATM x 100 MANs x 800 headends x 1000 households).

of more than 40 Mbps. Assuming an MPEG-2 movie of 3.35 Mbps (including video, audio, and systems), this will extend the capacity of the system by allowing more than ten MPEG-2 compressed movies to be transmitted in one 6 MHz channel (40 divided by 3.35).

Upgraded cable plant of 750 MHz and fiber-in-the-loop (FITL) technology will be able to serve 200 to 1000 households. The *500 channel scenario* will then consist of:

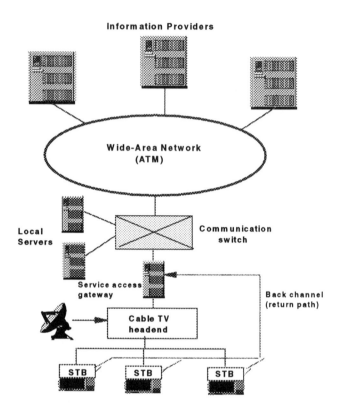

Figure 3.7   Cable architecture for interactive television [Mil94].

- about 70 analog channels (6 MHz channels, total approximately 450 MHz), and

- about 430+ digital and compressed channels (300 MHz/6 MHz = 50 analog channnels; each analog channel is capable of transmitting 8-10 MPEG-2 movies).

A major challenge for the cable companies is to install giant gateways, to lease backbone capacity from long distance carriers, or to lay out their own digital fiber trunk lines in order to construct a nation-wide network.

Example

In the following example consider the design of a local interactive television system, based on a local multimedia server linked to users via a metropolitan area network. The server serves the neighborhood of 1000 households. Assuming HDTV quality of videos, which requires 16 Mbps (or 2MBps), a 100 minute movie will require 12 GB of storage [VR93]. If there are 1000 different videos stored in the server, it will require 12 TB of storage, as illustrated in Figure 3.8. Disc technology will soon allow 100 GB per disc, so the server will require an array of 120 discs. Regarding transmission capacity, fiber-optic networks are already available which can offer tens of gigabits of bandwidth.

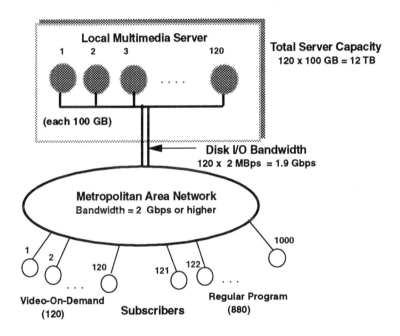

Figure 3.8 An example of a local multimedia server serving a neighborhood of 1000 households. The server provides HDTV-type movies on demand. Only 120 users can receive movies on-demand simultaneously.

In this example all 1000 users won't be able to receive movies-on-demand simultaneously. Assuming that disc channels can deliver 2 MBps per channel, 120 discs are capable of delivering 16 Mbps x 120 discs = 1.92 Gbps. Therefore, due to the I/O bottleneck, only 120 users will simultaneously receive HDTV movies-on-demand (1.92 Gbps/16 Mbps = 120). Other users can play videos or receive regular broadcast channels at that time.

To increase the number of users who can simultaneously receive movies-on-demand, the following can be done: (a) the number of discs can be increased, which will increase the I/O throughput, (b) instead 16 Mbps HDTV movies, 3-6 Mbps MPEG-2 movies can be transmitted, or (c) techniques for increasing the capacity of video-on-demand systems, such as segmentation and multicasting, can be deployed [NPS94, KF94].

For example, if the same system from Figure 3.8 is deployed to transmit 3 Mbps MPEG-2 movies, the number of users, who will receive the movies simultaneously, will increase to 640 (1.92 Gbps/3 Mbps = 640).

We briefly describe next segmentation and multicasting techniques for increasing the capacity of VOD systems.

The segmentation technique can overcome the problem of tying up network resources associated with continuous transmission of a video. Segmentation makes use of the high network rates and a limited buffer available at user's premises.

Assuming that delivery rate is thrice that of the playback rate, a 6 minute video can be transmitted in 2 minutes. This fact is used to divide the video into segments, as illustrated in Figure 3.9. The video in Figure 3.9 consists of 10 segments, and the segments are transmitted at regular intervals of 6 minutes. Each segment will be transmitted in 2 minutes, and therefore the whole 60-minute video will tie up a channel for only 20 minutes. The same channel can be used to serve other requests without disturbing the current session.

Multicasting is a communication technique in which data is sent to a group of users at the same time over a single channel. In a VOD system there is a chance that more than one user requests the same video at the same time. In this case, the video can be multicast, and all the users in the multicast group will be served by one channel thus saving the network bandwidth.

In the simplest case of multicasting, called synchronized multicasting, a video is multicast to users who request the video at the same time. When the video is segmented and users have the ability to buffer few segments of a video, multicasting is possible even for asynchronous requests for the same video [KF94].

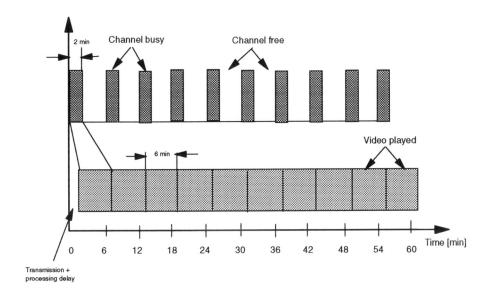

Figure 3.9 Example of the segmentation technique. A 60 min video is divided into 6 min segments and each segment is transmitted in 2 min.

Example of a commercial VOD ITV system. One of the first commercial video-on-demand ITV system, employed by Bell Atlantic, is targeted at entertainment, allowing residential subscribers to access a Video Information Provider's (VIP) database to view different movies. Other interactive services include home shopping and education. Bell Atlantic's system architecture is based on MPEG and ADSL (Asymmetric Digital Subscriber Line) technologies, as shown in Figure 3.10 [SL92].

A video gateway is used as a front-end to the system, providing customers with menus of available VIPs. Digital crossconnect systems facilitate switching connections between the VIPs and customers' loops. A voice switch is used to support POTS (Plain Old Telephone System) service. ADSL technology is deployed to deliver one channel of MPEG compressed movies to a single user. The data is sent through the switched network to a TV set top box that decompresses it and converts to NTSC analog video for delivery to TV. The system is under testing in northern Virginia serving about 2000 customers.

Another experimental VOD ITV system, based on SONET network, was developed at GTE Laboratories and is currently being tested in a trial in Cerritos,

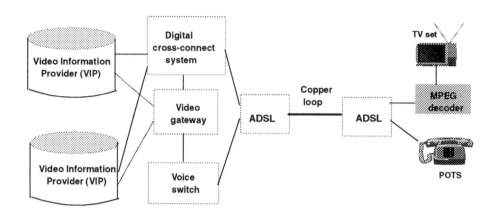

Figure 3.10   Bell Atlantic's architecture for VOD ITV services. ADSL is used to deliver one channel of MPEG compressed movies to a single user. Digital cross-connect system is used to switch connections between VIPs and customer loops.

California. The system deploys a large number of hierarchical switches, similar to those used in the telephone networks, to provide universal connectivity to an unlimited number of video service providers.

Interactive TV Set Top Box

A set top box (STB) for a full service network is a digital cable terminal device that allows users to interact with the network providing personalized, on-demand, interactive multimedia services [F+95]. The STB is the bridge between the user's display devices, peripherals, and input devices (such as infrared remote) on one side, and a digital and analog communication channel on the other side. This channel connects the STB to the information structure of service providers, as shown in Figures 3.5 and 3.6. In addition, the STB provides the capabilities that enable application developers to develop entertaining and compelling interfaces.

A typical STB should be capable of receiving a multi-modal data stream from the network, consisting of digital video, audio, graphics, text, and user-interface components. First generation STBs (1GSTB) include only decode/play functions and a simple reverse channel (e.g., Philips makes 1GSTB).

More sophisticated STBs should be capable of generating a multi-modal data stream comprising of the same components in order to transmit such data stream up through the network when sufficient bandwidth is available. However, throughout the initial phase of the information superhighway evolution, most of the data sent upstream will require low bandwidth as it will be of a low interactive nature and primarily for control purposes and ordering menus.

Second generation STBs (2GSTB) include local processors and support these additional capabilities. Many companies are developing 2GSTBs; among others Acorn, SGI, and Microsoft/GI.

A typical STB has two categories of physical interfaces: (a) to a digital/analog communication channel, which connects the STB to the information infrastructure, and (b) to the display devices (such as TV), VCR, remote controller, and various computer peripherals. Figure 3.11 shows TV STB interfaces.

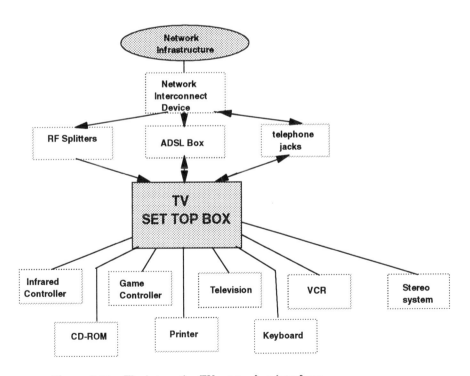

Figure 3.11   The interactive TV set top box interfaces.

Hardware Architecture of a STB. A potential hardware architecture of an interactive TV set top box, which is capable of supporting all the required functions, is shown in Figure 3.12 [F+95]. It consists of the following components:

- Network interface with upstream communication,

- Processing unit and memory,

- Peripheral control,

- Video subsystem,

- Graphics subsystem, and

- Audio subsystem.

The network interface connects the STB to the network infrastructure. It may incorporate the security services including permanent and renewable security. Some companies proposed digital delivery mechanisms (such as ATM) to the STB.

The processing unit runs a small real-time operating system to manage resources and activities in the STB. The system ROM contains boot code and basic OS services, while system DRAM memory is shared between operating system, applications, and data.

The peripheral control system enables a user to connect peripherals to the STB. The peripherals may include printers to print items such as coupons and programming guides, CD-ROM and magnetic disks for mass storage, digital video cameras, game controllers, and infrared controllers.

The video subsystem decompresses encoded video streams in formats such as MPEG-2. The processing of MPEG-2 video streams requires substantial processing power. Performance of the present processors still does not allow pure software MPEG decoder implementation, and therefore a hardware assistance is needed. A number of companies, including C-Cube, Philips, AT&T, LSI Logic, IBM, and SGS Thompson, have announced chip sets for MPEG-2 decoding.

The graphics subsystem is required for user interface elements for navigation and presentation. In addition, the graphics subsystem provides acceleration

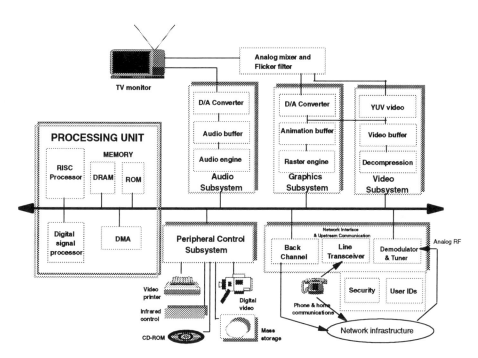

Figure 3.12 A potential hardware architecture of an interactive TV set top box.

for two- and three-dimensional graphics applications, such as interactive video games.

The audio subsystem decodes the audio stream corresponding to the video stream. The audio subsystem also generates audio locally in the form of MIDI and wave-table synthesis to accompany games and other applications.

In order to produce a STB at a low cost for the consumer market, integration of different blocks in the architecture will occur. A feasible approach is the integration of the major functions into four VLSI chips: processor, media processing, media access, and communication. The processor chip will include a powerful RISC processor, clock, and memory caches. The media processing chip will include the video decompression subsystem, a graphics accelerator, a display controller, color space conversion, and D/A output to the RF modulator. The media access chip will integrate the peripheral control functions. The communication chip will integrate all communication and network interface functions, such as demodulation and modulation, encryption and decryption, and security.

Software Architecture of a STB. A potential software architecture of the STB, shown in Figure 3.13, is layered to abstract device-dependent features [F+95]. The hardware Abstraction Layer (HAL) provides a low-level programmer's interface to different hardware subsystems, such as video, audio, network interface, and graphics subsystems.

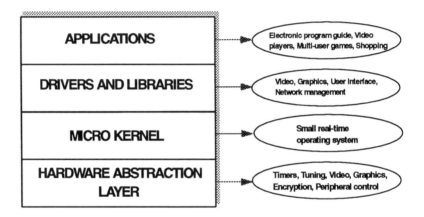

Figure 3.13   A potential software architecture of a set top box.

A small real-time operating system resides on the top of the HAL. The OS provides services such as process creation and execution, interprocess communications, and resource allocation and management. The resources managed by the OS include memory, communication channels, network bandwidth, and access to peripherals.

A suite of drivers and library routines provide common, frequently requested services to applications. These include Application Programmer's Interfaces (APIs) for network and session management, video control, graphics, and user interface.

Applications reside on the top of the software hierarchy. Examples of applications include electronic program guide, video player, home shopping, and various interactive games.

3.2.2 News On-Demand Distribution System

Another example of a multimedia on-demand system is the news production and delivery system. In this section we briefly describe a prototype news distribution system developed by Nynex and Dow Jones [MBG93]. The purpose of the system is to deliver business news on demand to various corporate and financial services.

The prototype system, shown in Figure 3.14, consists of the following major components:

- The Dow Jones Production Center,

- A wideband network,

- The Nynex Media Service Center, and

- Customers sites.

The stories are produced and published at the Production Center and then stored at several file servers. Then, the stored files are transferred through a wide area network to the multimedia storage subsystem at the Nynex Media Service Center. Customer sites can request specific news stories from the Media Service Center, which are then delivered over the wide area network and stored at customer sites.

The news stories consist of digitized audio and video files, which are stored on a audio/video file server. Text and graphics part is stored separately on a Novell file server.

The first implementation of the wide area network is based on T1 circuits; however the goal is to change it to a switched wideband service implemented on the primary-rate ISDN network. Each location in the network contains inverse multiplexers, which partition the available bandwidth on the T1 circuits into isochronous and packet data channels. The bandwidth of each T1 circuit is 1.344 Mbps. These devices, called bandwidth allocators, allocate channels depending whether there is a request for file transfer or live transmission.

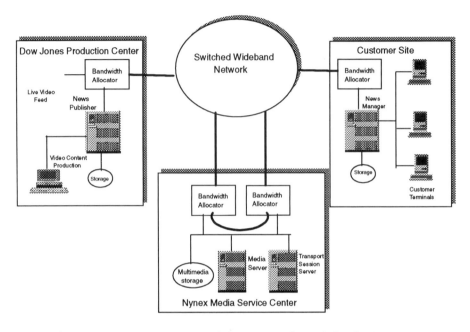

Figure 3.14 The architecture of the news on-demand distribution system
developed by Nynex and Dow Jones [MBG93].

3.3 RESEARCH DIRECTIONS

Research and development in high-speed networks will soon provide the bandwidth needed for distributed multimedia applications. Therefore, we envision
tremendous growth in distributed multimedia systems and their applications.
One of the major projects in high-speed networks is Blanca, part of the Gigabit
Testbed Initiative sponsored by the Corporation for National Research Initia-
tives and supported by the National Science Foundation. This project aims to
build two wide-area, ATM-based testbeds–Xunet 2 at 45 Mbps and the 622
Mbps Xunet 3–and to demonstrate on them several multimedia applications
that require high-speed networking.

Another nationwide initiative supported by the NSF is the development of
digital multimedia libraries. In this area, a working group for the Electronic
Library of the Future recently convened to conceive and develop the emerging
"library without walls." The first project involves the Library of Congress.

By the year 2000, the project hopes to have the core of the library's collection on-line and accessible over the networks.

Many challenging problems remain to be researched and resolved for the future growth of multimedia systems. Multimedia applications make enormous demands on computer hardware and software resources. Therefore, one of the ongoing demands is to develop more powerful multimedia workstations. Multimedia workstations will also need multimedia operating systems (MMOSs) and advanced multimedia user interfaces. A MMOS should handle continuous media by providing preemptive multitasking, easy expandability, format-independent data access, and support for real-time scheduling. It should be object-oriented and capable of synchronizing data streams. Users expect multimedia data to be instantly available, so the user interface must be highly sophisticated and intuitive. Integrating the user interface at the operating level could eliminate many problems for application software developers.

Other research challenges include developing new real-time compression algorithms (perhaps based on wavelets), video and image indexing and retrieval techniques, large storage devices, and multimedia data management systems. The constant challenge is further refinement of high-speed, deterministic networks with low latency and low jitter as well as research in new multimedia synchronization algorithms.

REFERENCES

[AE92]   S. R. Ahuja and J. R. Ensor, "Coordination and Control of Mul-
          timedia Conferencing", *IEEE Communications Magazine*, Vol. 30,
          No.5, May 1992, pp. 38-43.

[BG91]   L. O. Barbora and N. D. Georganas, "Multimedia Services and Ap-
          plications", *European Transactions on Telecommunications*, Vol. 2,
          No. 1, Jan/Feb 1991, pp. 5-19.

[BL91]   D. C. A. Bulterman and R. van Liere, "Multimedia Synchronization
          and Unix", *Proc. 2nd Int'l Workshop on Network and Operating
          System Support for Digital Audio and Video*, Heidelberg, Germany,
          November 1991, pp. 108-119.

[Bou92]   J. Y. L. Boudec, "The ATM: A Tutorial", *Computer Networks and
          ISDN Sytems*, Vol. 24, 1992, pp. 279-309.

[Buc94]   R. Buck, "The ORACLE Media Server for nCUBE Massively Par-
          allel Systems", *Proc. of the IEEE Int. Conf. on Parallel Processing*,
          Cancun, Mexico, April 1994, pp. 670-674.

[Cla92]   W. J. Clark, "Multiport Multimedia Conferencing", *IEEE Commu-
          nications*, Vol. 30, No. 5, May 1992, pp. 44-50.

[Col93]   B. Cole, "Interactive Multimedia: The Technology Framework",
          IEEE Spectrum, March 1993, pp. 32-39.

[Cri93]   M. Crimmins, "Analysis of Video Conferencing on a Token Ring
          Local Area Network", *Proc. ACM Multimedia 93*, ACM Press, New
          York, 1993, pp. 301-310.

[DCH94]   "Data Communication Handbook", *Siemens Stromberg-Carlson*,
          Boca Raton, Florida, 1994.

[F+95]   B. Furht, D. Kalra, A. Rodriguez, W. E. Wall, and F. Kitson, "De-
          sign Issues for Interactive Television Systems", *IEEE Computer*, Vol.
          28, No. 5, May 1995.

[FM95]   B. Furht and M. Milenkovic, "A Guided Tour of Multimedia Systems and Applications", IEEE Tutorial Text, *IEEE Computer Society Press*, 1995.

[Fur94]   B. Furht, "Multimedia Systems: An Overview", *IEEE Multimedia*, Vol. 1, No.1, Spring 1994, pp.47-59.

[Gon94]   F. Gong, "Multipoint Audio and Video Control for Packet-Based Multimedia Conferencing", *Proc. of the Second ACM Conference on Multimedia*, San Francisco, CA, 1994, pp. 425-432.

[HKS93]   M.J. Handley, P.T. Kirstein, and M.A. Sasse, "Multimedia Integrated Conferencing for European Researchers (MICE): Piloting Activities and the Conference management and Multiplexing Centre", *Computer Networks and ISDN Systems*, Vol. 26, 1993, pp. 275-290.

[IBM92]   "IBM Personal System/2 Multimedia Fundamentals", *IBM Corporation*, No. GG24-3653-01, June 1992.

[KF94]   H. Kalva and B. Furht, "Techniques for Improving the Capacity of Video-On-Deamnd Systems", *Technical Report TR-CSE-94-53*, Florida Atlantic University, Dept. of Computer Science and Engineering, Boca Raton, Florida, October 1994.

[KMF94]   H. Kalva, R. Mohammad, and B. Furht, "A Survey of Multimedia Networks", *Technical Report TR-CSE-94-42*, Florida Atlantic University, Dept. of Computer Science and Engineering, Boca Raton, Florida, September 1994.

[KP94]   J. Korst and V. Pronk, "Compact Disc Standards: An Introductory Overview", *ACM Journal on Multimedia Systems*, Vol. 2, No. 4, 1994, pp. 157-171.

[LG90]   T. D. C. Little and A. Ghafoor, "Network Considerations for Distributed Multimedia Object Composition and Communication", *IEEE Network*, Vol. 4, No. 6, Nov. 1990, pp. 32-49.

[LG91]   T. D. C. Little and A. Ghafoor, "Multimedia Synchronization Protocols for Broadband Integrated Services", *IEEE Journal on Selected Areas in Commmunications*, Vol. 9, No. 9, Dec. 1991, pp. 1,368-1,382.

[LV94]   T. D. C. Little and D. Venkatesh, "Prospects for Interactive Video-on-Demand", *IEEE Multimedia*, Vol. 1, No. 3, Fall 1994, pp. 14-24.

[MBG93]   G. Miller, G. Baber, and M. Gilliland, "News On-Demand for Multimedia Networks", *Proc. of the First ACM Conference on Multimedia*, Anaheim, CA, 1993, pp. 383-392.

[MDC92]   "Multimedia Distributed Computing", *IBM Corporation*, November 1992.

[Mil93]   M. Milenkovic, "Videoconferencing", *in Siggraph 93 Course Notes* on "Multimedia Systems: A Guided Tour", by M. Milenkovic and B. Furht, Anaheim, CA, August 1993.

[Mil94]   M. Milenkovic, "Multimedia Workstation and Interactive TV", *in ACM'94 Multimedia Tutorial Notes* on "Multimedia Systems and networks", by M. Milenkovic and B. Furht, San Francisco, CA, October 1994.

[MPM92]   "Multimedia Presentation Manager/2 - Programming Guide", *IBM Corporation*, 1992.

[NPS94]   J.-P. Nussbaumer, B.V. patel, and F. Schaffa, "Capacity Analysis of CATV for On-Demand Multimedia Distribution", *Proc. of the First ISMM Conf. on Distributed Multimedia Systems and Applications*, Honolulu, Hawaii, August 1994, pp. 97-100.

[PP93]   L.G. Palmer and R.S. Palmer, "DECspin: A Networked Desktop Videoconferencing Application", *Digital Technical Journal*, Vol. 5, No.3, Spring 1993, pp. 65-76.

[Pre93]   L. Press, "The Internet and Interactive Television", *Communications of the ACM*, Vol. 36, No. 12, December 1993, pp.19-23.

[R+91]   S. Ramanathan et al., "Optimal Communication Architecture for Multimedia Conferencing in Distributed Systems", *Tech. Report No. CS91-213*, University of California, San Diego, Computer Science and Engineering Department, October 1991.

[Rei94]   A. Reinhardt, "Building the Data Superhighway", *Byte*, March 1994. pp. 46-73.

[RKK94]   R.R. Roy, A.K. Kuthyar, and V. Katkar, "An Analysis of Universal Multimedia Switching Architectures", *AT&T Technical Journal*, Vol. 73, No. 6, November/December 1994, pp. 81.92.

[Roy94]   R.R. Roy, "Networking Constraints in Multimedia Conferencing and the Role of ATM Networks", *AT&T Technical Journal*, Vol. 73, No. 4, July/August 1994, pp. 97-108.

[RR94]   S. Ramanathan and P. Venkat Rangan, "System Architectures for Personalized Multimedia Services", *IEEE Multimedia*, Vol. 1, No. 1, Spring 1994, pp. 37-46.

[RVR92]   P. Venkat Rangan, H. M. Vin, and S. Ramanathan, "Designing an On-Demand Multimedia Service", *IEEE Communication Magazine*, July 1992, pp. 56-64.

[SL92]   J. Sutherland and L. Litteral, "Residential Video Services", *IEEE Communications Magazine*, July 1992, pp. 36-41.

[SVH91]   E. D. Sykas, K. M. Vlakos, and M. J. Hillyard, "Overview of ATM Networks: Functions and Procedures", *Computer Communications*, Vol. 14, No. 10, Dec. 1991, pp. 615-626.

[VR93]   H.M. Vin and R. Venkat Rangan, "Designing a Multi-User HDTV Storage Sever", *IEEE Journal on Selected Areas in Communications*, Vol. 11, No. 1, January 1993, pp. 153-164.

[Wil94]   L. Wilson, "Inside OS/2 Software Motion Video", *Dr. Dobb's Multimedia Sourcebook*, Vol. 18, No. 14, Winter 1994.

[WK90]   G. M. Woodruff and R. Kositpaiboon, "Multimedia Traffic Management Principles for Guaranteed ATM Network Performance", *IEEE Journal on Selected Areas in Communications*, Vol. 8, No. 3, April 1990, pp. 437-446.

QUESTIONS AND PROBLEMS - PART I

1. Name and briefly describe the key elements of a multimedia workstation.

2. Compare various network topologies and protocols of the candidate systems for the information superhighway.

3. Which (synchronous or asynchronous) mode of communication is better suited for multimedia traffic and why?

4. Which problem is addressed by implementing priority in token ring networks? State briefly how it works and what benefits it brings to priority traffic.

5. Describe four data location models related to multimedia synchronization. Draw simple diagrams, and briefly discuss synchronization issues related to these four models.

6. Describe the synchronization algorithm implemented in the IBM's Multimedia Presentation Manager/2 system.

7. Discuss and compare features of various switching technologies (ATM, STM, SMDS, Frame Relay) for multimedia communications.

8. (a) Draw a hierarchical configuration of multimedia servers and network switches needed for on-demand multimedia services. Briefly describe its components.

   (b) For a residential neighborhood of 2,000 households, calculate the number of discs (N) needed for storage and the bandwidth of the ATM switch (B), if:

   - each client requires independent selection of 1 hour video,
   - videos are MPEG-2 encoded, requiring 4 Mbits/sec.
   - available discs are of 12 Gbytes capacity.

9. A remote database contains 60 seconds of motion-video, with 30 frames/sec. Each frame, stored in a compressed form, requires 3 Mbits of storage. This motion-video is transmitted via the communication channel with the capacity of 60 Mbits/sec, as shown in figure below.

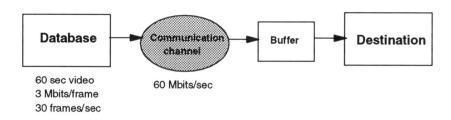

a. Calculate the total number of frames (F) and the storage requirements (Sr) in Mbytes for this motion-video.

b. Does the required motion video transfer (Rt) exceed the communication channel capacity? If this is true, explain how you would resolve this problem, and:

• calculate the delay time at the receiving site (Tdelay),

• show the timing diagram which will include packet production schedule and playout schedule.

c. Calculate the required size of the buffer (Bs) at the receiving site in order to achieve the proposed playout schedule.

10. Explain principles of operation of Asynchronous Transfer Mode (ATM). Specifically address the following issues:

• primary intended purpose,
• types of traffic intended to carry,
• mode of operation - asynchronous, circuit or packet switched,
• types of adapation layers.

11. In a distributed multimedia system define the end-to-end delay in the case of continuous synchronization. Draw the timing diagram which includes all the times. What is the required end-to-end delay for the videophone application:

(a) 0.05sec   (b) 0.25 sec   (c) 1 sec

12. Explain the difference between continuous and point synchronization of multimedia streams. Use diagrams.

13. Show the timing diagram of two streams (audio and video) and define speed ratio, jitter, and skew.

14. Discuss advantages and disadvantages of various network configurations for videoconferencing applications.

15. What are hardware and software components of a multimedia workstation for videoconferencing?

16. What is the required one-way end-to-end delay for videoconferencing applications? Specify all time components that comprise the total time.

17. A metropolitan server is serving a neighborhood of 2000 households. Assume that the server transmits 5 Mbps MPEG-2 movies; each movie is of 2 hours duration.

    (a) How many discs are needed at the server site to store 2000 different movies? Assume that a disc capacity is 100 GB.

    (b) Assume that disc channels can deliver 2 MBps. How many users can simultaneously receive on-demand movies? The network's bandwidth is 3 Gbps.

    (c) Describe techniques for increasing the number of users who will simultaneously receive the movies.

18. Explain how a "500 channel scenario" can be achieved using both analog and digital TV channels.

19. What are main services of an interactive television system?

PART II
Multimedia Compression Techniques and Standards

4

INTRODUCTION TO MULTIMEDIA COMPRESSION

4.1 STORAGE REQUIREMENTS FOR MULTIMEDIA APPLICATIONS

Audio, image, and video signals require a vast amount of data for their representation. Table 4.1 illustrates the mass storage requirements for various media types, such as text, image, audio, animation, and video.

There are three main reasons why present multimedia systems require that data must be compressed. These reasons are related to:

- Large storage requirements of multimedia data,

- Relatively slow storage devices which do not allow playing back uncompressed multimedia data (specifically video) in real time, and

- The present network's bandwidth, which does not allow real-time video data transmission.

To illustrate large storage requirements, consider a typical multimedia application, such as encyclopedia, which may require:

- 500,000 pages of text (2 KB per page) - total 1 GB,

- 3,000 color pictures (in average 640x480x24 bits = 1 MB/picture) - total 3 GB,

	TEXT	IMAGE	AUDIO	ANIMATION	VIDEO
OBJECT TYPE	-ASCII -EBCDIC	-Bit-mapped graphics -Still photos -Faxes	Non-coded stream of digitized audio or voice	Synched image and audio stream at 15-19 frames/s	TV analog or digital image with synched streams at 24-30 frames/s
SIZE AND BANDWIDTH	2 KB per page	-Simple (grayscale) **77KB** per image (320x240x8bits) -Detailed (color) **3 MB** per image (1100x900x24bits)	Voice/Phone 8KHz/8 bits- (mono) 6-44 KB/s Audio CD 44.1 KHz, 16 bits/stereo **176 KB/s** (44.1KHz x 2 ch x 16 bits)	16 bit color, 16 frames/sec **6.5 MB/s** (32 x640 x16bits x 16 frames/sec)	24 bit color, 30 frames/sec **27.6 MB/s** (640X480X24bits x 30 frames/sec)

Table 4.1   Storage requirements for various media types.

- 500 maps (in average 640x480x16 bits = 0.6 MB/map) - total 0.3 GB,

- 60 minutes of stereo sound (176 KB/second) - total 0.6 GB,

- 30 animations, in average 2 minutes in duration (640x320x16 bits x 16 frames/sec = 6.5 MB/second) - total 23.4 GB,

- 50 digitized movies, in average 1 minute in duration (640x480x24 bits x 30 frames/sec = 27.6 MB/sec) - total 82.8 GB.

The encyclopedia will require total of 111.1 GB storage capacity. Consider that compression algorithms are then applied to compress various media used in the encyclopedia. Assume that the following average compression ratios are obtained:

- Text 2:1,

- Color images 15:1,

- Maps 10:1,

- Stereo sound 6:1,

- Animation 50:1,

- Motion video 50:1.

Figure 4.1 gives the storage requirements for the encyclopedia before and after compression. When using compression, storage requirements will be reduced from 111.1 GB to only 2.96 GB, which is much easier to handle.

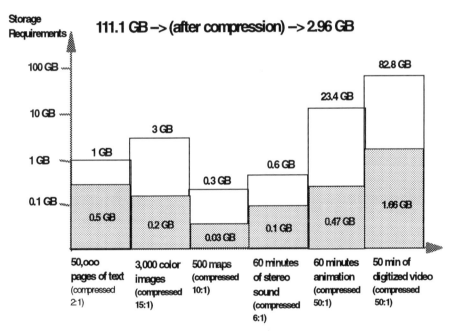

Figure 4.1 Storage requirements for the encyclopedia before and after compression.

To illustrate the second reason for compression of multimedia data, which is slow transmission rates of current storage devices, consider playing a one-hour movie from a storage device, such as a CD-ROM. Assuming color video frames with a resolution of 620x560 pixels and 24 bits/pixel, it would require about 1 MB/frame of storage. For a motion video requiring 30 frames/second, it gives a total of 30 MB for one second of motion video, or 108 GB for the whole one-hour movie. Even if there is enough storage capacity available, we won't be able to play the video in real time due to insufficient transmission rates of current storage devices.

According to the previous calculation, the required transmission rate of the storage device should be 30 MB/sec; however today's technology provides about 300 KB/sec transfer rate of CD-ROMs. Therefore, a one-hour movie would be played for 100 hours! In summary, at the present state of technology of storage devices, the only solution is to compress multimedia data before its storage and decompress it before the playback.

The final reason for compression of multimedia data is the limited bandwidth of present communication networks. The bandwidth of traditional networks (Ethernet, token ring) is in tens of Mb/sec, which is too low even for the transfer of only one motion video in uncompressed form. The newer networks, such as ATM and FDDI, offer a higher bandwidth (in hundreds of Mb/sec to several Gb/sec), but only few simultaneous multimedia sessions would be possible if the data is transmitted in uncompressed form.

Modern image and video compression techniques offer a solution of this problem, which reduces these tremendous storage requirements. Advanced compression techniques can compress a typical image ranging from 10:1 to 50:1. Very high compression ratios of up to 2000:1 can be achieved in compressing of video signals.

4.2   CLASSIFICATION OF COMPRESSION TECHNIQUES

Compression of digital data is based on various computational algorithms, which can be implemented either in software or in hardware. Compression techniques are classified into two categories: (a) lossless, and (b) lossy approaches [Fox91, Fur94]. Lossless techniques are capable to recover the original representation perfectly. Lossy techniques involve algorithms which recover the presentation to be similar to the original one. The lossy techniques provide higher compression ratios, and therefore they are more often applied in image and video compression than lossless techniques. The classification schemes for lossless and lossy compression are presented in Figures 4.2a and 4.2b, respectively.

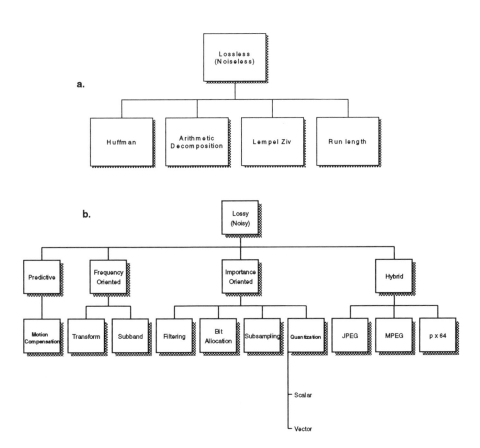

Figure 4.2 The classification of compression techniques: (a) Lossless techniques, (b) Lossy techniques.

The lossy techniques are classified into: (1) prediction based techniques, (2) frequency oriented techniques, and (3) importance-oriented techniques. Prediction based techniques, such as ADPCM, predict subsequent values by observing previous values. Frequency oriented techniques apply the Discrete Cosine Transform (DCT), which relates to fast Fourier transform. Importance-oriented techniques use other characteristics of images as the basis for compression. For example, DVI technique uses color lookup tables and data filtering.

The hybrid compression techniques, such as JPEG, MPEG, and px64, combine several approaches, such as DCT and vector quantization or differential pulse code modulation. Recently, standards for digital multimedia have been established based on these three techniques, as illustrated in Table 4.2.

Short Name	Official Name	Standards Group	Compression Ratios
JPEG	Digital compression and coding of continuous-tone still images	Joint Photographic Experts Group	15:1 Full color still-frame applications
H.261 or px64	Video encoder /decoder for audio-visual services at px64 Kbps	Specialist Group on Coding for Visual Telephony	100:1 to 2000:1 Video-based tele-communications
MPEG	Coding of moving pictures and associated audio	Moving Pictures Experts Group	200:1 Motion-intensive applications

Table 4.2　Multimedia compression standards.

4.3　IMAGE CONCEPTS AND STRUCTURES

A digital image represents a two-dimensional array of samples, where each sample is called a pixel. Precision determines how many levels of intensity can be represented, and is expressed as the number of bits/sample. According to precision, images can be classified into:

Binary images, represented by 1 bit/sample. Examples include black/white photographs and facsimile images.

Computer graphics, represented by a lower-precision, as 4 bits/sample.

Grayscale images, represented by 8 bits/sample.

Color images, represented with 16, 24 or more bits/sample.

According to the trichromatic theory , the sensation of color is produced by selectively exciting three classes of receptors in the eye. In a RGB color representation system, shown in Figure 4.3, a color is produced by adding three primary colors: red, green and blue (RGB). The straight line, where $R = G = B$, specifies the gray values ranging from black to white.

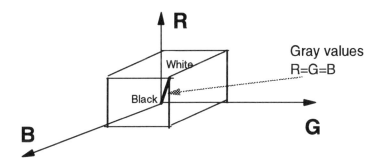

Figure 4.3 The RGB representation of color images.

Figure 4.4 illustrates how a three-sensor RGB color video camera operates and produces colors at a RGB monitor [Poy94]. Lights for sources of different colors are added together to produce the prescribed color.

Another representation of color images, YUV representation, describes luminance and chrominance components of an image. The luminance component provides a grayscale version of the image, while two chrominance components give additional information that converts the grayscale image to a color image.

The YUV representation is more natural for image and video compression. The exact RGB to YUV transformation , defined by the CCIR 601 standard , is given by the following transformations:

$$Y = 0.299R + 0.587G + 0.114B \tag{4.1}$$

$$U = 0.564(B - Y) \tag{4.2}$$

$$V = 0.713(B - Y) \tag{4.3}$$

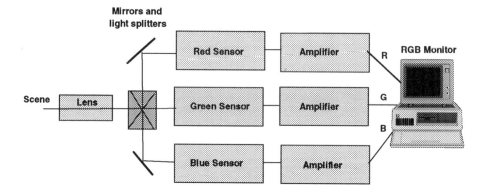

Figure 4.4   The operation of a three-sensor RGB color video camera.

where Y is the luminance component, and U and V are two chrominance components.

An approximate RGB to YUV transformation is given as:

$$Y = 0.3R + 0.6G + 0.1B \qquad (4.4)$$

$$U = B - Y \qquad (4.5)$$

$$V = R - Y \qquad (4.6)$$

This transformation has a nice feature that, when $R = G = B$, then $Y = R = G = B$, and $U = V = 0$. Then, this represents a grayscale image.

Color conversation RGB to YUV requires several multiplication operations, which can be computationally expensive. An approximation, proposed in [W+94], can be calculated by performing bit shifts and adds instead multiplication operations. This approximation is given as:

$$Y = \frac{R}{4} + \frac{G}{2} + \frac{B}{2} \qquad (4.7)$$

$$U = \frac{B - Y}{2} \qquad (4.8)$$

$$V = \frac{R - Y}{2} \qquad (4.9)$$

This approximation also gives a simplified YUV to RGB transformation, expressed by:

$$R = Y + 2V \qquad (4.10)$$
$$G = Y - (U + V) \qquad (4.11)$$
$$B = Y + 2U \qquad (4.12)$$

Another color format, referred to as YCbCr format , is intentively used for image compression. In YCbCr format, Y is the same as in a YUV system, however U and V components are scaled and zero-shifted to produce Cb and Cr, respectively, as follows:

$$C_b = \frac{U}{2} + 0.5 \qquad (4.13)$$

$$C_r = \frac{V}{1.6} + 0.5 \qquad (4.14)$$

In this way, chrominance components Cb and Cr are always in the range [0,1].

Resolutions of an image system refers to its capability to reproduce fine detail. Higher resolution requires more complex imaging systems to represent these images in real time. In computer systems, resolution is characterized with number of pixels (for example, VGA has a resolution of 640 x 480 pixels). In video systems, resolution refers to the number of line pairs resolved on the face of the display screen, expressed in cycles per picture height, or cycles per picture width. For example, the NTSC broadcast system in North America and Japan, denoted 525/59.94, has about 483 picture lines (525 denotes the total number of lines in its rates, and 59.94 is its field rate in Hertz). The HDTV system will approximately double the number of lines of current broadcast television at approximately the same field rate. For example, a 1152x900 HDTV system may have 937 total lines and a frame rate of 65.95 Hz.

Figure 4.5 compares various image structures, showing the vertical and horizontal pixel counts, and the approximate total number of pixels [LH91, Poy94].

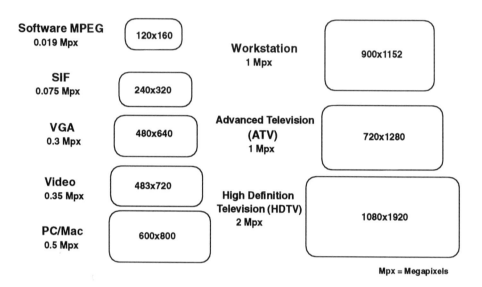

Figure 4.5   Various image structures.

The CCITT has adopted two picture formats for video-based telecommunications, Common Intermediate Format (CIF) , and Quarter-CIF (QCIF), described in detail in Section 6.1.

The full-motion video is characterized with at least 24 Hz frame rate (or 24 frames/sec), and up to 30, or even 60 frames/sec for HDTV. For animation, acceptable frame rate is in the range 15-19 frames/sec, while for video telephony is 5-10 frames/sec. Videoconferencing and interactive multimedia applications require a rate of 15-30 frames/sec.

5

JPEG ALGORITHM FOR FULL-COLOR STILL IMAGE COMPRESSION

Originally, JPEG standard was targeted for full-color still frame applications, achieving 15:1 average compression ratio [PM93, Wal91, Fur95]. However, JPEG has also been applied in some real-time, full-motion video applications (Motion JPEG - MJPEG). JPEG standard provides four modes of operation:

- *sequential DCT-based encoding*, in which each image component is encoded in a single left-to-right, top-to-bottom scan,

- *progressive DCT-based encoding*, in which the image is encoded in multiple scans, in order to produce a quick, rough decoded image when the transmission time is long,

- *lossless encoding*, in which the image is encoded to guarantee the exact reproduction, and

- *hierarchical encoding*, in which the image is encoded in multiple resolutions.

5.1 JPEG CODEC

In this section we describe both the design of a sequential JPEG encoder and decoder. The block diagrams of the JPEG sequential encoder and decoder are shown in Figure 5.1.

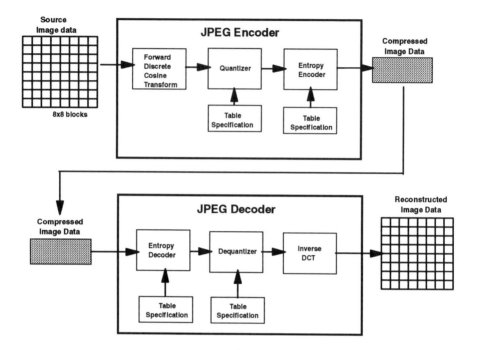

Figure 5.1   Block diagrams of sequential JPEG encoder and decoder.

5.1.1   JPEG Encoder

The JPEG encoder, shown in Figure 5.1, consists of three main blocks:

- Forward Discrete Cosine Transform (FDCT) block,

- Quantizer, and

- Entropy encoder.

At the input of the encoder, the original unsigned samples, which are in the range $[0, 2^p - 1]$, are shifted to signed integers with range $[-2^{p-1}, 2^{p-1} - 1]$. For example, for a grayscale image, where p=8, the original samples with range [0, 255], are shifted to range [-128, +127].

Then, the source image is divided into 8x8 blocks, and samples from each block are transformed into the frequency domain using the Forward Discrete Cosine Transform using the following equations:

$$F(u, v) = \frac{C(u)}{2} \cdot \frac{C(v)}{2} \sum_{x=0}^{7} \sum_{y=0}^{7} f(x, y) \cos\frac{(2x + 1)u\pi}{16} \cos\frac{(2y + 1)v\pi}{16} \quad (5.1)$$

where:

$$C(u) = \frac{1}{\sqrt{2}} \quad \text{for} \quad u = 0$$

$$C(u) = 1 \quad \text{for} \quad u > 0$$

$$C(v) = \frac{1}{\sqrt{2}} \quad \text{for} \quad v = 0$$

$$C(v) = 1 \quad \text{for} \quad v > 0$$

The transformed 64-point discrete signal is a function of two spatial dimensions x and y, and its components are called spatial frequencies or DCT coefficients.

The $F(0, 0)$ coefficient is called the "DC coefficient", and the remaining 63 coeffcents are called the "AC coefficients". For a grayscale image, the obtained DCT coefficients are in the range [-1024, +1023], which requires additional 3 bits for their representation, compared to the original image samples. Several fast DCT algorithms are proposed and analysed in [PM93, HM94].

For a typical 8x8 image block, most spatial frequencies have zero or near-zero values, and need not to be encoded. This is illustrated in the JPEG example, presented in Section 5.2. This fact is the foundation for achieving data compression.

In the next block, quantizer, all 64 DCT coefficients are quantized using a 64-element quantization table, specified by the application. The quantization reduces the amplitude of the coefficients which contribute little or nothing to the quality of the image, with the purpose of increasing the number of zero-value coefficients. Quantization also discards information which is not visually significant. The quantization is performed according to the following equation:

$$F_q(u, v) = Round \left[\frac{F(u, v)}{Q(u, v)}\right] \qquad (5.2)$$

where Q(u,v) are quantization coefficients specified by a quantization table. Each element Q(u,v) is an integer from 1 to 255, which specifies the step size of the quantizer for its corresponding DCT coefficient.

A set of four quantization tables are specified by the JPEG standard for compliance testing of generic encoders and decoders; they are given in Table 5.1. In the JPEG example in Section 5.2, a quantization formula is used to produce quantization tables.

After quantization, the 63 AC coefficients are ordered into the "zig-zag" sequence, as shown in Figure 5.2. This zig-zag ordering will help to facilitate the next phase, entropy encoding, by placing low-frequency coefficients, which are more likely to be nonzero, before high-frequency coefficients. This fact is confirmed by the experiment presented in [PM93]; obtained results are given in Figure 5.3. These results show that, when the coefficients are ordered zig-zag, the probability of coefficients being zero is an increasing monotonic function of the index. The DC coefficients, which represent an average value of the 64 image samples, are encoded using the predictive coding techniques, as illustrated in Figure 5.4.

The reasons for predictive coding of DC coefficients is that there is usually a strong correlation between DC coefficients of adjacent 8x8 blocks. Adjacent blocks will very probably have similar average intensities. Therefore, coding the differences between DC coefficients rather than the coefficients themselves will give better compression.

Finally, the last block in the JPEG encoder is the entropy coding, which provides additional compression by encoding the quantized DCT coefficients into more compact form. The JPEG standard specifies two entropy coding

8	6	5	8	12	20	26	30
6	6	7	10	13	29	30	28
7	7	8	12	20	29	35	28
7	9	11	15	26	44	40	31
9	11	19	28	34	55	52	39
12	18	28	32	41	52	57	46
25	32	39	44	52	61	60	51
36	46	48	49	56	50	52	50

9	9	12	24	50	50	50	50
9	11	13	33	50	50	50	50
12	13	28	50	50	50	50	50
24	33	50	50	50	50	50	50
50	50	50	50	50	50	50	50
50	50	50	50	50	50	50	50
50	50	50	50	50	50	50	50
50	50	50	50	50	50	50	50

16	17	18	19	20	21	22	23
17	18	19	20	21	22	23	24
18	19	20	21	22	23	24	25
19	20	21	22	23	24	25	26
20	21	22	23	24	25	26	27
21	22	23	24	25	26	27	28
22	23	24	25	26	27	28	29
23	24	25	26	27	28	29	30

16	16	19	22	26	27	29	34
16	16	22	24	27	29	34	37
19	22	26	27	29	34	34	38
22	22	26	27	29	34	37	40
22	26	27	29	32	35	40	48
26	27	29	32	35	40	48	58
26	27	29	34	38	46	56	69
27	29	35	38	46	56	69	83

Table 5.1   Four quantization tables for compliance testing of generic JPEG encoders and decoders.

methods: Huffman coding and arithmetic coding . The baseline sequential JPEG encoder uses Huffman coding, which is presented next.

The Huffman encoder converts the DCT coefficients after quantization into a compact binary sequence using two steps: (1) forming intermediate symbol sequence, and (2) converting intermediate symbol sequence into binary sequence using Huffman tables .

In the intermediate symbol sequence, each AC coefficient is represented by a pair of symbols:

- Symbol-1 (RUNLENGTH,SIZE), and
- Symbol-2 (AMPLITUDE).

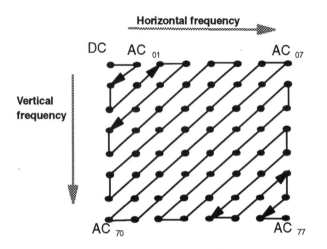

Figure 5.2   Zig-zag ordering of AC coefficients.

RUNLENGTH is the number of consecutive zero-valued AC coefficients preceding the nonzero AC coefficient. The value of RUNLENGTH is in the range 0 to 15, which requires 4 bits for its representation.

SIZE is the number of bits used to encode AMPLITUDE. The number of of bits for AMPLITUDE is in the range of 0 to 10 bits, so there are 4 bits needed to code SIZE.

AMPLITUDE is the amplitude of the nonzero AC coefficient in the range of [+1024 to -1023], which requires 10 bits for its coding. For example, if the sequence of AC coefficients is:

$$\underbrace{0,0,0,0,0,0,}_{6} \ 476$$

the symbol representation of the AC coefficient 476 is:

$$(6,9) \, (476)$$

where: RUNLENGTH=6, SIZE=9, and AMPLITUDE=476.

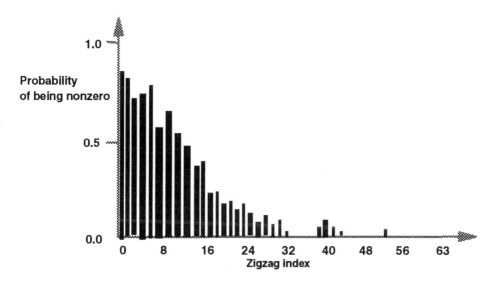

Figure 5.3   The probability of being nonzero of zig-zag ordered AC coefficients.

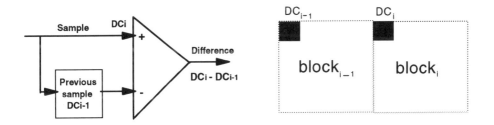

Figure 5.4   Predictive coding of DC coefficients. The difference between the present and the previous DC coefficients is calculated and then coded using JPEG.

If RUNLENGTH is greater than 15, then Symbol-1 (15,0) is interpreted as the extension symbol with RUNLENGTH=16. These can be up to three consecutive (15,0) extensions.

In the following example:

$$(15, 0)\,(15, 0)\,(7, 4)\,(12)$$

RUNLENGTH is equal to 16+16+7=39, SIZE=4, and AMPLITUDE=12.

The symbol (0,0) means 'End of block' (EOB) and terminates each 8x8 block.

For DC coefficients, the intermediate symbol representation consists of:

- Symbol-1 (SIZE), and
- Symbol-2 (AMPLITUDE).

Because DC coefficients are differentially encoded, this range is double the range of AC coefficients, and is [-2048, +2047].

The second step in Huffman coding is converting the intermediate symbol sequence into a binary sequence. In this phase, symbols are replaced with variable length codes, beginning with the DC coefficient, and continuing with the AC coefficients .

Each Symbol-1 (both for DC and AC coefficients) is encoded with a Variable-Length Code (VLC), obtained from the Huffman table set specified for each image component. The generation of Huffman tables is discussed in [PM93]. Symbols-2 are encoded using a Variable-Length Integer (VLI) code, whose length in bits is given in Table 5.2.

For example, for an AC coefficeint presented as the symbols:

$$(1, 4)\,(12)$$

the binary presentation will be: (1 1 1 1 1 0 1 1 0 1 1 0 0), where (1 1 1 1 1 0 1 1 0) is VLC obtained from the Huffman table, and (1 1 0 0) is VLI code for 12.

5.1.2   JPEG Decoder

In the JPEG sequential decoding, all the steps from the encoding process are inversed and implemented in reverse order, as shown in Figure 5.1. First, an

Size	Amplitude range
1	(-1,1)
2	(-3,-2) (2,3)
3	(-7..-4) (4..7)
4	(-15..-8) (8..15)
5	(-31..-16) (16..31)
6	(-63..-32) (32..63)
7	(-127..-64) (64..127)
8	(-255..-128) (128..255)
9	(-511..-256) (256..511)
10	(-1023..-512) (512..1023)

Table 5.2   Huffman coding of Symbols-2.

entropy decoder (such as Huffman) is implemented on the compressed image data. The binary sequence is converted to a symbol sequence using Huffman tables (VLC coefficients) and VLI decoding, and then the symbols are converted into DCT coefficients. Then, the dequantization is implemented using the following function:

$$F_q'(u, v) = F_q(u, v) \times Q(u, v) \tag{5.3}$$

where Q(u,v) are quantization coefficients obtained from the quantization table.

Then, the Inverse Discrete Cosine Transform (IDCT) is implemented on dequantized coefficients to convert the image from frequency domain into spatial domain. The IDCT equation is defined as:

$$F(x, y) = \frac{1}{4}[\sum_{u=0}^{7}\sum_{v=0}^{7}C(u)C(v)F(u, v)cos\frac{(2x + 1)u\pi}{16}cos\frac{(2y + 1)v\pi}{16}] \tag{5.4}$$

where:

$$C(u) = \frac{1}{\sqrt{2}} \quad \text{for} \quad u = 0$$

$$C(u) = 1 \quad \text{for} \quad u > 0$$

$$C(v) = \frac{1}{\sqrt{2}} \quad \text{for} \quad v = 0$$

$$C(v) = 1 \quad \text{for} \quad v > 0$$

The last step consists of shifting back the decompressed samples in the range $[0, 2^p - 1]$.

5.2   COMPRESSION MEASURES

The basic measure for the performance of a compression algorithm is Compression Ratio (C_R), defined as:

$$C_R = \frac{\text{Original data size}}{\text{Compressed data size}} \tag{5.5}$$

There is a trade-off between the compression ratio and the picture quality. Higher compression ratios will produce lower picture quality and vice versa. Quality and compression can also vary according to source image characteristics and scene content. A measure for the quality of the picture, proposed in [Wal95], is the number of bits per pixel in the compressed image (N_b). This measure is defined as the total number of bits in the compressed image divided by the number of pixels:

$$N_b = \frac{\text{Encoded number of bits}}{\text{Number of pixels}} \tag{5.6}$$

According to this measure, four different picture qualities are defined [Wal91], as shown in Table 5.3.

Nb [bits/pixel]	Picture Quality
0.25 - 0.5	Moderate to good quality
0.5 - 0.75	Good to very good quality
0.75 - 1.0	Excellent quality
1.5 - 2.0	Usually indistinguishable from the original

Table 5.3   Picture quality characteristics.

Another statistical measure, that can be used to evaluate various compression algorithms, is the Root Mean Square (RMS) error, calculated as:

$$RMS = \frac{1}{n}\sqrt{\sum_{i=1}^{n}(X_i - \hat{X}_i)^2} \qquad (5.7)$$

where:

X_i - original pixel values,

$\hat{X}_i$ - pixel values after decompression,

n - total number of pixels in an image.

The RMS shows the statistical difference between the original and decompressed images. In most cases the quality of a decompressed image is better with lower RMS. However, in some cases it may happen that the quality of a decompressed image with higher RMS is better than one with lower RMS.

In the next section, we calculate these measures in several examples.

5.3 SEQUENTIAL JPEG ENCODING EXAMPLES

5.3.1 Encoding a Single Block

To illustrate all steps in baseline sequential JPEG encoding, we present the step-by-step procedure and obtained results in encoding an 8x8 block of 8-bit samples, as illustrated in Figure 5.5. The original 8x8 block is shown in Figure 5.5a; this block after shifting is given in Figure 5.5b. After applying the FDCT, the obtained DCT coefficients are given in Figure 5.5c. Note that, except for low-frequency coefficents, all other coefficients are close to zero.

For the generation of quantization tables, we used the program proposed in [Nel92]:

$$for(i = 0; i < N; i++)$$

$$for(j = 0; j < N; j++)$$

$$Q[i][j] = 1 + [(1 + i + j) \times quality];$$

The parameter *quality* specifies the quality factor, and its recommended range is from 1 to 25. *Quality* $= 1$ gives the best quality, but the lowest compression ratio, and *quality* $= 25$ gives the worst quality and the highest compression ratio. In this example, we used *quality* $= 2$, which generates the quantization table shown in Figure 5.5d.

After implementing quantization, the obtained quantized coefficients are shown in Figure 5.5e. Note that a large number of high-frequency AC coefficients are equal to zero.

The zig-zag ordered sequence of quantized coefficients is shown in Figure 5.5f, and the intermediate symbol sequence in Figure 5.5g. Finally, after implement-

(a) Original 8x8 block

```
140 144 147 140 140 155 179 175
144 152 140 147 140 148 167 179
152 155 136 167 163 162 152 172
168 145 156 160 152 155 136 160
162 148 156 148 140 136 147 162
147 167 140 155 155 140 136 162
136 156 123 167 162 144 140 147
148 155 136 155 152 147 147 136
```

(b) Shifted block

```
12 16 19 12 11 27 51 47
16 24 12 19 12 20 39 51
24 27  8 39 35 34 24 44
40 17 28 32 24 27  8 32
34 20 28 20 12  8 19 34
19 39 12 27 27 12  8 34
 8 28 -5 39 34 16 12 19
20 27  8 27 24 19 19  8
```

(c) Block after FDCT Eq. (5.1)

```
185 -17 14  -8  23 -9 -13 -18
 20 -34 26  -9 -10 10  13   6
-10 -23 -1   6 -18  3 -20   0
 -8  -5 14 -14  -8 -2  -3   8
 -3   9  7   1 -11 17  18  15
  3  -2 -18  8   8 -3   0  -6
  8   0 -2   3  -1 -7  -1  -1
  0  -7 -2   1   1  4  -6   0
```

(d) Quantization Table (quality=2)

```
 3  5  7  9 11 13 15 17
 5  7  9 11 13 15 17 19
 7  9 11 13 15 17 19 21
 9 11 13 15 17 19 21 23
11 13 15 17 19 21 23 25
13 15 17 19 21 23 25 27
15 17 19 21 23 25 27 29
17 19 21 23 25 27 29 31
```

(e) Block after quantization Eq. (5.2)

```
61 -3  2  0  2  0  0 -1
 4 -4  2  0  0  0  0  0
-1 -2  0  0 -1  0 -1  0
 0  0  1  0  0  0  0  0
 0  0  0  0  0  0  0  0
 0  0 -1  0  0  0  0  0
 0  0  0  0  0  0  0  0
 0  0  0  0  0  0  0  0
```

(f) Zig-zag sequence

61,-3,4,-1,-4,2,0,2,-2,0,0,0,0,0,2,0,0,0,1,0,0,0,0,0,0,-1,0,0,-1,0,0,
0,0,-1,0,0,0,0,0,0,-1,0,0,0,0,0,0,0,0,0,0,0,0,0,0,0,0,0,0,0,0,0,0,0

(g) Intermediate symbol sequence

(6)(61),(0,2)(-3),(0,3)(4),(0,1)(-1),(0,3)(-4),(0,2)(2),(1,2)(2),(0,2)(-2),
(0,2)(-2),(5,2)(2),(3,1)(1),(6,1)(-1),(2,1)(-1),(4,1)(-1),(7,1)(-1),(0,0)

(e) Encoded bit sequence (total 98 bits)

(110)(111101) (01)(00) (100)(100) (00)(0) (100)(001) (01)(10)
(11011)(10) (01)(01) (01)(01) (11111110111)(10) (111010)(1)
(1111011)(0) (11100)(0) (111011)(0) (11111010)(0) (1010)

Figure 5.5   Step-by-step procedure in JPEG sequential encoding of a 8x8 block.

ing Huffman codes, the obtained encoded bit sequence is shown in Figure 5.5h. The Huffman table used in this example is proposed in the JPEG standard for luminance AC coefficients [PM93], and the partial table, needed to code the symbols from Figure 5.5g, is given in Table 5.4.

(RUNLENGTH, SIZE)	Code Word
(0,0) EOB	1010
(0,1)	00
(0,2)	01
(0,3)	100
(1,2)	11011
(2,1)	11100
(3,1)	111010
(4,1)	111011
(5,2)	11111110111
(6,1)	1111011
(7,1)	11111010

Table 5.4   Partial Huffman table for luminance AC coefficients.

Note that the DC coefficient is treated as being from the first 8x8 block in the image, and therefore it is coded directly (not using predictive coding as all the remaining DC coefficients).

For this block, the compression ratio is calculated as:

$$C_R = \frac{\text{Original number of bits}}{\text{Encoded number of bits}} = \frac{64 \times 8}{98} = \frac{512}{98} = 5.22$$

and the number of bits/pixel in the compressed form is:

$$N_b = \frac{\text{Encoded number of bits}}{\text{Number of pixels}} = \frac{98}{64} = 1.53$$

5.3.2   Encoding Grayscale Images

In this section, we present results of experiments obtained by implementing the sequential JPEG algorithm for grayscale images [KAF94]. The JPEG algorithm was implemented and run on two Sun machines: Sparc 10 Model 41, characterized by 109 MIPS and 22 MFLOPS, and Sparc IPC, characterized by 17 MIPS and 2.1 MFLOPS. We used the algorithm, described in 5.2, to generate various quantization tables, selecting the following values for the parameter *quality*: 1, 2, 4, 8, 16, and 25. Grayscale images (320 x 200 pixels, 8 bits/pixel) were compressed and decompressed using the JPEG algorithm.

Complete results are given in [KAF94]; here results for the grayscale image 'Lisa' are presented. The empirical results are given in Table 5.5, and the original and decompressed images, for different quality factors, are shown in Figure 5.6.

Quality factor	Original number of bits	Compressed number of bits	Compression ratio (Cr)	Bits/pixel (Nb)	RMS error	Execution times [ms]	
						SUN SPARC 10/41	SUN IPC
1	512,000	48,021	10.66	0.75	2.25	0.59	6.31
2	512,000	30,490	16.79	0.48	2.75	0.59	6.22
4	512,000	20,264	25.27	0.32	3.43	0.58	6.39
8	512.000	14,162	36.14	0.22	4.24	0.59	6.44
15	512,000	10,479	48.85	0.16	5.36	0.58	6.45
25	512,000	9,034	56.64	0.14	6.40	0.58	6.32
DC only	512,000	7,688	66.60	0.12	7.92	0.57	6.25

Table 5.5   Results of JPEG compression of grayscale image 'Lisa' (320x240 pixels).

Depending on the selected quality, the obtained compression ratio varies from about 10 (quality factor=1) to about 66 (only the DC coefficient was used). In all experiments higher compression ratio produced lower quality images. Consequently, with a higher compression ratio the number of bits/pixel decreases and RMS error increases. Execution times, which include both encoder and decoder times, are almost identical for different quality levels. An average pro-

(a) Original image

(b) Quality=1, Cr=10.66

(c) Quality=2, Cr=16.79

(d) Quality=15, Cr=48.85

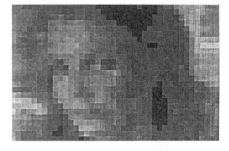

(e) DC coefficient only, Cr=66.60

Figure 5.6   Original image 'Lisa' and after decompression for different quality factors.

cessing speed at Sparc 10 has been over six times faster than at the Sparc IPC, which is consistent with the reported MIPS performance of these two machines.

5.4 JPEG COMPRESSION OF COLOR IMAGES

5.4.1 Generalizing to Multiple Components

The described sequential JPEG algorithm can be easily expanded for compression of color images, or in a general case for compression of multiple-component images. The JPEG source image model consists of 1 to 255 image components [Wal91, Ste94], called color or spectral bands, as illustrated in Figure 5.7.

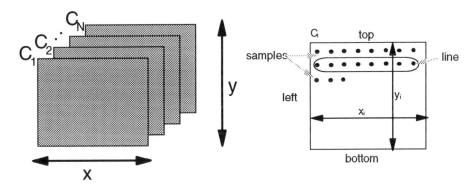

Figure 5.7 JPEG color image model.

For example, both RGB and YUV representations consist of three color components. Each component may have a different number of pixels in the horizontal (X_i) and vertical (Y_i) axis. Figure 5.8 illustrates two cases of a color image with 3 components. In the first case, all three components have the same resolutions, while in the second case they have different resolutions.

The color components can be processed in two ways:

(a) *Non-interleaved data ordering,* in which processing is performed component by component from left-to-right and top-to-bottom. In this mode,

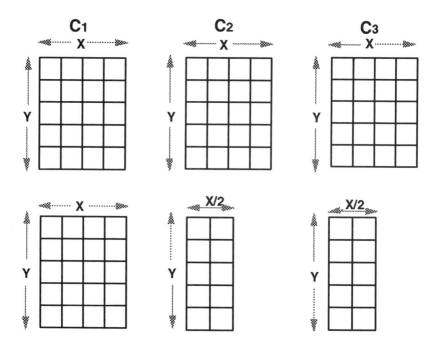

Figure 5.8 A color image with 3 components: (a) with same resolutions, (b) with different resolutions.

for a RGB image with high resolution, the red component will be displayed first, then the green component, and finally the blue component. Non-interleaved data ordering is illustrated in Figure 5.9.

(b) *Interleaved data ordering*, in which different components are combined into so-called Minimum Coded Units (MCUs).

Interleaved data ordering is used for applications that need to display or print multiple-component images in parallel with their decompression. When image components are interleaved, each component is partitioned into rectangular regions of H_i and V_i data units. Figure 5.10 illustrates an example of an image consisting of four components C_1, C_2, C_3 and C_4, each having different $\{H_i, V_i\}$. The JPEG algorithm specifies the Minimum Coded Unit (MCU), as the smallest group of interleaved data units, as shown in Figure 5.10.

Block diagrams of the encoder and decoder for color JPEG compression are identical to those for grayscale image compression, shown in Figure 5.1, except

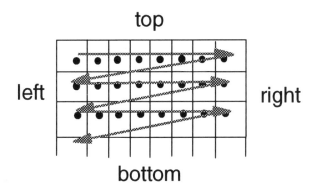

Figure 5.9   Noninterleaved data ordering in JPEG compression of multiple-components images.

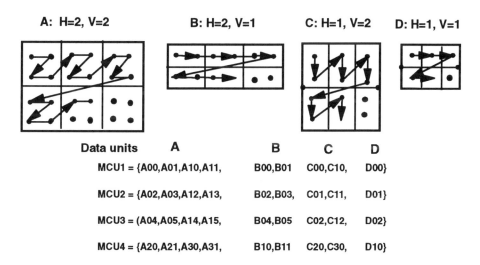

Figure 5.10   Interleaved data ordering in JPEG compression of color images. The MCU consists of data units taken from each component A, B, C, and D.

the first block into encoder is a color space conversion block (for example, RGB to YUV conversion), and at the decoder side the last block is the inversed color conversion, such as YUV to RGB.

5.4.2   Experimental Data

In this experiment, the execution time of a color JPEG algorithm was analysed [MF94a]. The goal was to measure execution times of different blocks in the JPEG encoder and decoder and identify blocks take more time. Originally, this experiment was performed in order to parallelize some of the steps in JPEG algorithm and design the most effective parallel JPEG algorithm.

The color JPEG algorithm was run on an i486/33MHz machine with no maths coprocessor. The 320x240 'Targa' image, with 24 bits/pixel, was compressed and then decompressed. The results are presented in Table 5.6.

OPERATION	JPEG COMPRESSION		JPEG DECOMPRESSION	
	Time [sec]	Percentage	Time [sec]	Percentage
Read/Write File	0.75	11	0.34	6
RGB/YUV Conversion	1.40	21	1.14	19
Reorder from/for YUV	0.28	4	0.49	8
FDCT/IDCT	2.87	43	2.90	48
Quantization/ Dequantization	0.47	7	0.37	6
Huffman enc./dec.	0.87	13	0.70	12
Write/Read File I/O	0.01	1	0.06	1
TOTAL TIME	**6.65**	**100%**	**6.00**	**100%**

Table 5.6   Experimental data: JPEG compression of a color image.

From Table 5.6 it can be seen that the JPEG algorithm spends 48% of decompression time performing IDCT, and similarly 42% of compression time performing FDCT. Color conversion requires about 20% and Huffman coding about 12% of the total execution time.

Assuming a Pentium personal computer with math coprocessor, the expected performance of the JPEG algorithm can be improved about sixfold, which will give about 1 sec for decompression.

5.5   PROGRESSIVE JPEG COMPRESSION

5.5.1   A Multiple Pass Approach

In some applications an image may have large numbers of pixels and the decompression process, including transmission of the compressed image over the network, may take several minutes. In such applications, there may be a need to produce quickly a rough image, and then improve its quality using multiple scans [PM93, Wal91, Ste94]. The progressive JPEG mode of operation produces a sequence of scans, each scan coding a subset of DCT coefficients. Therefore, the progressive JPEG encoder must have an additional buffer at the output of the quantizer and before the entropy encoder. The size of the buffer should be large enough to store all DCT coefficients of the image, each of which is 3 bits larger than the original image samples.

Figure 5.11 illustrates the differences in displaying a decompressed image in the progressive and sequential JPEG.

Progressive

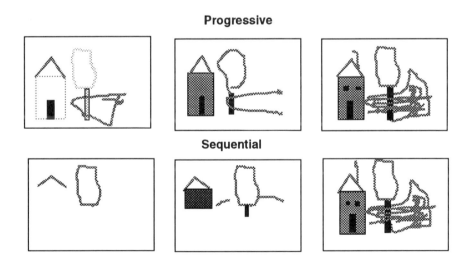

Sequential

Figure 5.11   Progressive versus sequential JPEG encoding.

Progressive JPEG compression can be achieved using three algorithms:

(1) Progressive spectral selection algorithm,

(2) Progressive successive approximation algorithm,

(3) Combined progressive algorithm.

In the *progressive spectral selection algorithm*, the DCT coefficients are grouped into several spectral bands. Typically, low-frequency DCT coefficients bands are sent first, and then higher-frequency coefficients. For example, a sequence of four spectral bands may look like this (see Figure 5.12):

band 1:   DC coefficient only

band 2: AC_1 and AC_2 coefficients

band 3: AC_3, AC_4, AC_5, AC_6 coefficients

band 4: $AC_7....AC_{63}$ coefficients

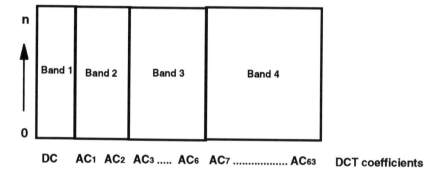

Figure 5.12   An example of applying the spectral selection progressive JPEG algorithm using four scans.

In the *progressive successive approximation algorithm*, all DCT coefficients are sent first with lower precision, and then refined in later scans. For example, a sequence of three successive approximation bands may be as follows (see Figure 5.13):

band 1:   All DCT coefficients divided by 4

band 2:   All DCT coefficients divided by 2

band 3:   All DCT coefficients at full resolution

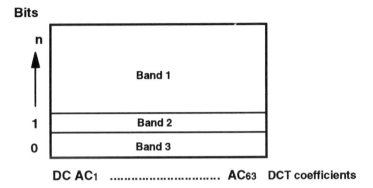

Figure 5.13   An example of applying the successive approximation progressive JPEG algorithm using three scans.

Combined progressive algorithm combines both spectral selection and successive approximation algorithms. Figure 5.14 illustrates an image divided into 8 combined scans. For example, in the first scan only DC coefficients divided by 2 (with lower resolution) will be sent, then AC band-1 will be sent, and so on.

5.5.2   Experimental Data

In this experiment, we implemented both progressive algorithms, spectral selection (SS), and successive approximation (SA) to compress and decompress a 320x200 grayscale image 'Cheetah'. In both cases, we used four scans, as shown in Table 5.7.

The results of the experiments are presented in Tables 5.8 and 5.9. Note that in both algorithms, the first scan will be transmitted 6-7 times faster than using the sequential JPEG algorithm (26,215 bits in SA and 29,005 bits in SS versus 172,117 bits in sequential JPEG).

The total number of transmitted bits, which includes all four scans, is about 230,000 bits for spectral selection, and about 220,000 for the successive approximation technique, which is slightly more than in the case of sequential JPEG (172,000 bits).

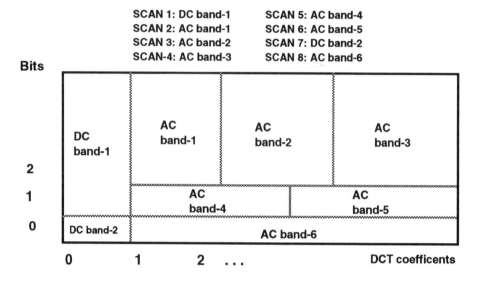

Figure 5.14 An example of applying combined progressive JPEG algorithm consisting of eight scans.

	Spectral Seelection	Successive Approximation
SCAN 1	DC, AC1, AC2	All DCT - divided by 8
SCAN 2	AC3 - AC9	All DCT - divided by 4
SCAN 3	AC10 - AC35	All DCT divided by 2
SCAN 4	AC 36 - AC 63	All DCT - full resolution

Table 5.7 Four scans applied in the progressive JPEG experiments.

Figures 5.15 and 5.16 show results of sequential and progressive JPEG compression using EasyTech/Codec, respectively. In the case of sequential JPEG, obtained compression ratios are in the range 21 to 52. Assuming an image transmitted over a 64 Kbits/s ISDN network, the following four images are produced using progressive JPEG (see Figure 5.16). The first image of reduced quality appears on the screen very quickly, in 0.9 seconds. Each subsequent pass improves the image quality and is obtained after 1.6, 3.6, and 7.0 seconds, respectively.

Scan number	Bits transmitted	Compression ratio	Bits/pixel	RMS error
1	29,005	17.65	0.45	19.97
2	37,237	7.73	1.04	13.67
3	71,259	3.72	2.15	7.90
4	92,489	3.01	2.66	4.59
Sequential JPEG	172,117	2.97	2.69	4.59

Table 5.8 Progressive spectral selection JPEG. Image 'Cheetah': 320x240 pixels = 512,000 bits.

Scan number	Bits transmitted	Compression ratio	Bits/pixel	RMS error
1	26,215	19.53	0.41	22.48
2	34,506	8.43	0.95	12.75
3	63,792	4.11	1.95	7.56
4	95,267	2.33	2.43	4.59
Sequential JPEG	172,117	2.97	2.69	4.59

Table 5.9 Progressive successive approximation JPEG. Image 'Cheetah': 320x240 pixels = 512,000 bits.

ORIGINAL

COMPRESSION FACTOR 21:1 (low)

COMPRESSION FACTOR 33:1 (medium)

COMPRESSION FACTOR 52:1 (high)

Figure 5.15   Sequential JPEG results using Easy/Tech Codec. (Courtesy of AutoGraph International.)

IMAGE AFTER 0.9 SECONDS

IMAGE AFTER 1.6 SECONDS

IMAGE AFTER 3.6 SECONDS

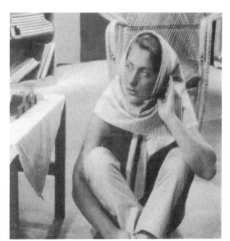

IMAGE AFTER 7.0 SECONDS

Figure 5.16   Progressive JPEG results using Easy/Tech Codec. (Courtesy of AutoGraph International.)

5.6 SEQUENTIAL LOSSLESS JPEG COMPRESSION

The JPEG standard also supports a lossless mode of operation, by providing a simple predictive compression algorithm, rather than DCT-based technique, which is a lossy one. Figure 5.17 shows the block diagram of the lossless JPEG encoder, in which a prediction block has replaced the FDCT and the quantization blocks from the baseline sequential DCT-based JPEG encoder.

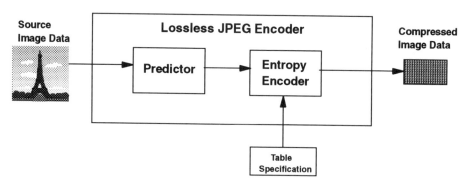

Figure 5.17   Block diagram of the lossless JPEG encoder.

The predictor block works in such a way that a prediction of the sample $\hat{X}$ is calculated on the basis of previous samples A, B and C, and then the difference $\Delta X = X - \hat{X}$ is computed, where X is the actual value of the sample (Figure 5.18a and b). Then, the difference ΔX is encoded using the Huffman encoder.

Table 5.10 illustrates several different predictor formulas that can be used for the lossless prediction. Lossless JPEG compression typically gives around 2:1 compression ratio for moderately complex color images.

5.7 HIERARCHICAL JPEG COMPRESSION

Hierarchical JPEG compression offers a progressive representation of a decoded image similar to progressive JPEG, but also provides encoded images at multiple resolutions. Hierarchical JPEG mode of operation creates a set of com-

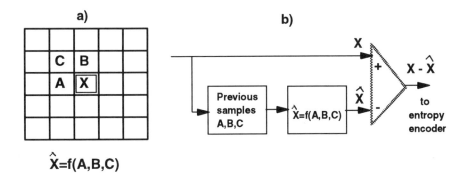

Figure 5.18 Lossless JPEG encoding: (a) Locations of four samples in the predictor, (b) Predictor block diagram.

Selection Value	Predictor Formula
0	no prediction
1	X = A
2	X = B
3	X = C
4	X = A+B-C
5	X = A + (B-C)/2
6	X = B+ (A-C)/2
7	X = (A+B)/2

Table 5.10 Predictors for lossless JPEG compression.

pressed images beginning with small images, and then continuing with images with increased resolutions. This process is called downsampling, or pyramidal coding [PM93, Wal91, Ste94], as illustrated in Figure 5.19.

After the downsampling phase, each lower-resolution image is scaled up to the next resolution (upsampling process), and used as a prediction for the following

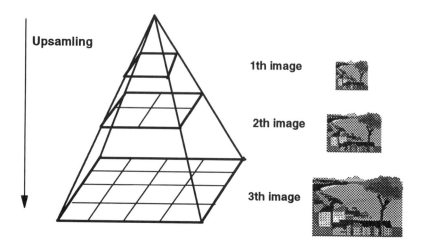

Figure 5.19 Hierarchical JPEG encoding consists of multiple-resolution images.

stage. The hierarchical JPEG compression algorithm consists of the following steps:

- Filter and downsample the original image by the desired number of multiples of 2 in each dimension.

- Encode the reduced-size image using either sequential DCT, progressive DCT, or lossless encoders.

- Decode this reduced-size image and then interpolate and upsample it by 2, horizontally and/or vertically, using the identical interpolation filter which the receiver will use.

- Use the obtained upsampled image as a prediction of the original, and encode the image differences using one of the encoders.

- Go to step 1 until the full resolution is encoded.

Hierarchical JPEG encoder requires significantly more buffer space, but the benefits are that the encoded image is immediately available at different resolutions. Several models for hierarchical coding, such as DCT-based and lossless correction models, are discussed in [PM93].

6

PX64 COMPRESSION ALGORITHM FOR VIDEO TELECOMMUNICATIONS

The H.261 standard, commonly called px64, is optimized to achieve very high compression ratios for full-color, real-time motion video transmission. The px64 compression algorithm combines intraframe and interframe coding to provide fast processing for on-the-fly video compression and decompression. *Intraframe coding* refers to the coding of individual frames, while *interframe coding* is the coding of a frame in reference to the previous or future frames.

The px64 standard is optimized for applications such as video-based telecommunications. Because these applications are usually not motion-intensive, the algorithm uses limited motion search and estimation strategies to achieve higher compression ratios. For standard video communication images, compression ratios of 100:1 to over 2000:1 can be achieved.

The px64 compression standard is intended to cover the entire ISDN channel capacity (p = 1, 2, ...30). For p = 1 to 2, due to limited available bandwidth, only desktop face-to-face visual communications (videophone) can be implemented using this compression algorithm. However, for $p \geq 6$, more complex pictures are transmitted, and the algorithm is suitable for videoconferencing applications.

6.1   CCITT VIDEO FORMAT

The px64 algorithm operates with two picture formats adopted by the CCITT, Common Intermediate Format (CIF), and Quarter-CIF (QCIF) [Lio91], as illustrated in Table 6.1.

	CIF		QCIF	
	Lines/frame	Pixels/line	Lines/frame	Pixels/line
Luminance (Y)	288	352	144	176
Chrominance (Cb)	144	176	72	88
Chrominance (Cr)	144	176	72	88

Table 6.1 Parameters of the CCITT video formats.

Intended applications of this standard are videophone and videoconferencing applications. The following examples illustrates the need for a video compression algorithm for these applications.

Example 1: Desktop videophone application

For a desktop videophone application, if we assume p=1, the available ISDN network bandwidth is $B_A = 64$ Kbits/sec. If the QCIF format is used, the required number of bits per frame consists of one luminance and two chrominance components:

$$N_b = (144 \times 176 + 72 \times 88 + 72 \times 88) \times 8 \text{ bits} = 300 \text{ Kbits/frame}$$

If the data is transmitted at 10 frames/sec, the required bandwidth is:

$$B_r = 300 \text{ Kbits/frame} \times 10 \text{ frames/sec} = 3 \text{ Mbits/sec}$$

As a consequence, a video compression algorithm should provide compression ratio of minimum:

$$C_r = \frac{B_r}{B_A} = \frac{3 \text{ Mbits/sec}}{64 \text{ Kbits/sec}} = 47$$

Example 2: Videoconferencing application

Assuming p=10 for a videoconferencing application, the available ISDN network bandwidth becomes $B_r = 640$ Kbits/sec. If the CIF format is used, the total number of bits per frame becomes:

$$N_b = (288 \times 352 + 144 \times 176 + 144 \times 176) \times 8 \text{ bits} = 1.21 \text{ Mbits/frame}$$

Assuming a frame rate of 30 frames/sec, the required bandwidth for the transmission of videoconferencing data becomes:

$$B_r = 1.21 \text{ Mbits/frame} \times 30 \text{ frames/sec} = 36.4 \text{ Mbits/sec}$$

Therefore, a video compression algorithm should provide compression ratio of minimum:

$$C_r = \frac{B_r}{B_A} = \frac{36.4 \text{ Mbits/sec}}{640 \text{ Kbits/sec}} = 57$$

6.2   PX64 ENCODER AND DECODER

6.2.1   Algorithm Structure

The px64 video compression algorithm combines intraframe and interframe coding to provide fast processing for on-the-fly video. The algorithm consists of:

- DCT-based intraframe compression, which similarly to JPEG, uses DCT, quantization and entropy coding, and

- Predictive interframe coding based on Differential Pulse Code Modulation (DPCM) and motion estimation.

The block diagram of the px64 encoder is presented in Figure 6.1.

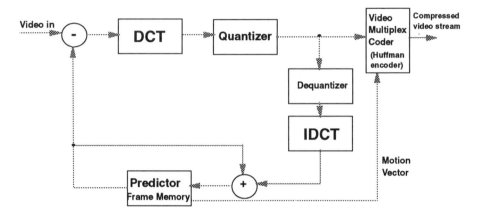

Figure 6.1   Block diagram of the px64 encoder.

The algorithm begins by coding an intraframe block using the DCT transform coding and quantization (intraframe coding) , and then sends it to the video multiplex coder. The same frame is then decompressed using the inverse quantizer and IDCT, and then stored in the picture memory for interframe coding.

During the interframe coding, the prediction based on the DPCM algorithm is used to compare every macro block of the actual frame with the available macro blocks of the previous frame, as illustrated in Figure 6.2. To reduce the encoding delay, only the closest previous frame is used for prediction.

Then, the difference is created as error terms, DCT-coded and quantized, and sent to the video multiplex coder with or without the motion vector. At the final step, entropy coding (such as Huffman encoder) is used to produce more compact code. For interframe coding, the frames are encoded using one of the following three techniques:

- DPCM coding with no motion compensation (zero-motion vectors),

- DPCM coding with non-zero motion vectors,

- Blocks are filtered by an optional predefined filter to remove high-frequency noise.

At least one in every 132 picture frames should be intraframe coded.

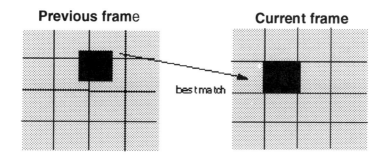

Figure 6.2 The principle of interframe coding in the px64 video compression.

A typical px64 decoder, shown in Figure 6.3, consists of the receiver buffer, the Hufffman decoder, inverse quantizer, IDCT block, and the motion-compensation predictor which includes frame memory.

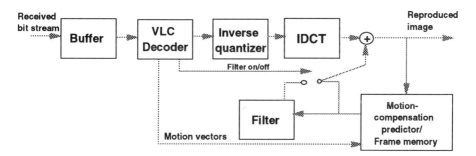

Figure 6.3 Block diagram of the px64 decoder.

The motion estimation algorithms are discussed in Chapter 7 on MPEG algorithm.

6.2.2   Video Data Structure

According to the H.261,standard, a data stream has a hierarchical structure consisting of Pictures, Groups of Blocks (GOB), Macro Blocks (MB) and Blocks [Lio91, A+93b]. A Macro Block is composed of four (8 x 8) luminance (Y) blocks, and two (8 x 8) chrominance (C_r and C_b) blocks, as illustrated in Figure 6.4.

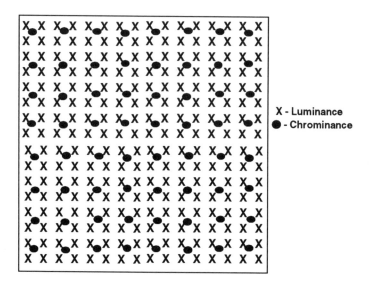

Figure 6.4   The composition of a Macro Block: MB = 4Y + Cb + Cr.

A Group of Blocks is composed of 3 x 11 MBs. A CIF Picture contains 12 GOBs, while a QCIF Picture consists of 4 GOBs. The hierarchical block structure is shown in Figure 6.5.

Each of the layers contains headers, which carry information about the data that follows. For example, a picture header includes a 20-bit picture start code, video format (CIF or QCIF), frame number, etc. A detailed structure of the headers is given in [Lio91].

A new H.261 codec, proposed in [Gha92], expands the existing H.261 codec to operate in ATM networks. A software-based video compression algorithm, called the Popular Video Codec (PVC), proposed in [HHW93], is suitable for real-time systems. The PVC coder simplifies compression and decompression processes of the px64 algorithm by removing the transform and the motion estimation parts, and modifies the quantizer and the entropy coder.

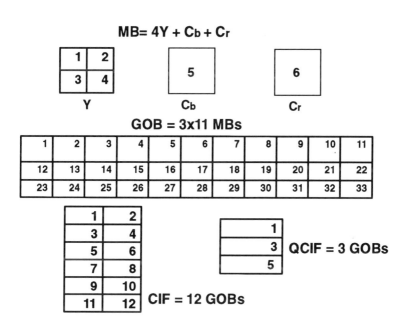

Figure 6.5 Hierarchical block structure of the px64 data stream.

7

MPEG COMPRESSION FOR MOTION-INTENSIVE APPLICATIONS

The MPEG compression algorithm is intended for compression of full-motion video. The compression method uses interframe compression and can achieve compression ratios of 200:1 through storing only the differences between successive frames. The MPEG approach is optimized for motion-intensive video applications, and its specification also includes an algorithm for the compression of audio data at ratios ranging from 5:1 to 10:1.

The MPEG first-phase standard (MPEG-1) is targeted for compression of 320×240 full-motion video at rates of 1 to 1.5 Mb/s in applications, such as interactive multimedia and broadcast television. MPEG-2 standard is intended for higher resolutions, similar to the digital video studio standard CCIR 601 , EDTV, and further leading to HDTV. If specifies compressed bit streams for high-quality digital video at the rate of 2-80 Mb/s. The MPEG-2 standard supports interlaced video formats and a number of features for HDTV. The MPEG-2 standard is also addressing scalable video coding for a variety of applications which need different image resolutions, such as video communications over ISDN networks using ATM [Ste94, LeG91]. The MPEG-4 standard is intended for compression of full-motion video consisting of small frames and requiring slow refreshments. The data rate required is 9-40 Kbps, and the target applications include interactive multimedia and video telephony. This standard requires the development of new model-based image coding techniques for human interaction and low-bit-rate speech coding techniques [Ste94].

Table 7.1 illustrates various motion-video formats and corresponding MPEG parameters.

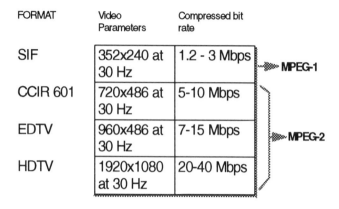

Table 7.1   Parameters of MPEG standards.

The MPEG algorithm is intended for both asymmetric and symmetric applications. Asymmetric applications are characterized by frequent use of the decompression process, while the compression process is performed once. Examples include movies-on-demand, electronic publishing, and education and training. Symmetric applications require equal use of the compression and decompression processes. Examples include multimedia mail and videoconferencing.

When MPEG standard is conceived, the following features have been identified as important: random access, fast forward/reverse searches, reverse playback, audio-visual synchronization, robustness to errors, editability, format flexibility, and cost tradeoff. These features are described in detail in [LeG91].

The MPEG standard consists of three parts: (1) synchronization and multiplexing of video and audio, (2) video, and (3) audio. In this chapter we describe the algorithms for MPEG video and audio compression (Sections 7.1 and 7.2), and the creation of the interleaved MPEG video/audio stream (Section 7.3). Experimental data are presented in Section 7.4.

7.1 MPEG VIDEO ENCODER AND DECODER

7.1.1 Frame Structures

In the MPEG standard, frames in a sequence are coded using three different algorithms, as illustrated in Figure 7.1.

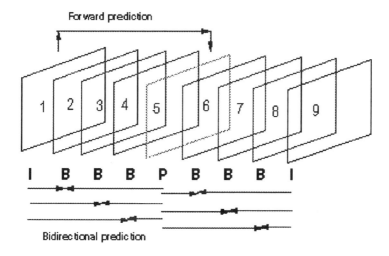

Figure 7.1 Types of frames in the MPEG standard.

I frames (intra images) are self-contained and coded using a DCT-based technique similar to JPEG. **I** frames are used as random access points in MPEG streams, and they give the lowest compression ratios within MPEG.

P frames (predicted images) are coded using forward predictive coding, where the actual frame is coded with reference to a previous frame (**I** or **P**). This process is similar to H.261 predictive coding, except the previous frame is not always the closest previous frame, like in H.261 coding (see Figure 6.2). The compression ratio of **P** frames is significantly higher than of **I** frames.

B frames (bidirectional or interpolated images) are coded using two reference frames, a past and a future frame (which can be **I** or **P** frames). Bidirectional, or interpolated coding provides the highest amount of compression.

Note that in Figure 7.1, the first 3 **B** frames (2, 3 and 4) are bidirectionally coded using the past frame **I** (frame 1), and the future frame **P** (frame 5). Therefore, the decoding order will differ from the encoding order. The **P** frame 5 must be decoded before **B** frames 2, 3 and 4, and **I** frame 9 before **B** frames 6, 7 and 8. If the MPEG sequence is transmitted over the network, the actual transmission order should be {1, 5, 2, 3, 4, 9, 6, 7, 8}.

The MPEG application determines a sequence of **I**, **P** and **B** frames. If there is a need for fast random access, the best resolution would be achieved by coding the whole sequence as **I** frames (MPEG becomes identical to MJPEG). However, the highest compression ratio can be achieved by incorporating a large number of **B** frames. The following sequence has been proven very effective for a number of practical applications [Ste94]:

$$(I\,B\,B\,P\,B\,B\,P\,B\,B)\ (I\,B\,B\,P\,B\,B\,P\,B\,B)...$$

In the case of 25 frames/s, random access will be provided through 9 still frames (**I** and **P** frames), which is about 360 ms [Ste94]. On the other hand, this sequence will allow a relatively high compression ratio.

7.1.2 Motion Estimation

The coding process for **P** and **B** frames includes the motion estimator, which finds the best matching block in the available reference frames. **P** frames are always using forward prediction, while **B** frames are using bidirectional prediction, also called motion-compensated interpolation, as illustrated in Figure 7.2 [A+93b].

B frames can use forward, backward prediction, or interpolation. A block in the current frame (**B** frame) can be predicted by another block from the past reference frame (**B** = **A** → forward prediction), or from the future reference frame (**B** = **C** → backward prediction), or by the average of two blocks (**B** = (**A** + **C**)/2 → interpolation).

Motion estimation is used to extract the motion information from the video sequence. For every 16 × 16 block of **P** and **B** frames, one or two motion vectors are calculated. One motion vector is calculated for **P** and forward and backward predicted **B** frames, while two motion vectors are calculated for interpolated **B** frames.

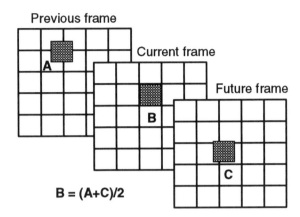

Figure 7.2 Motion compensated interpolation implemented in MPEG. Each block in the current frame is interpolated using the blocks from a previous and a future frame.

The MPEG standard does not specify the motion estimation technique, however block-matching techniques are likely to be used. In block-matching techniques, the goal is to estimate the motion of a block of size $(n \times m)$ in the present frame in the relation to the pixels of the previous or the future frames. The block is compared with a corresponding block within a search area of size $(m + 2p \times n + 2p)$ in the previous (or the future) frame, as illustrated in Figure 7.3a. In a typical MPEG system, a match block (or a macroblock) is 16×16 pixels $(n = m = 16)$, and the parameter p=6 (Figure 7.3b).

Many block matching techniques for motion vector estimation have been developed and evaluated in the literature, such as:

1. The exhaustive search (or brute force) algorithm,

2. The three-step-search algorithm,

3. The 2-D logarithmic search algorithm,

4. The conjugate direction search algorithm,

5. The parallel hierarchical 1-D search algorithm, and

6. The modified pixel-difference classification, layered structure algorithm.

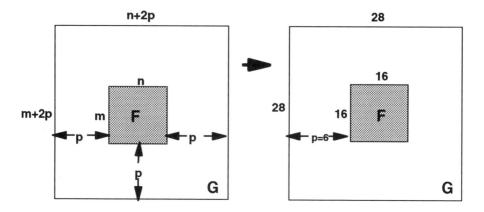

Figure 7.3 The search area in block-matching techniques for motion vector estimation: (a) a general case, (b) a typical case for MPEG: n=m=16, p=6. F - a macroblock in the current frame, G - search area in a previous (or a future) frame.

Cost Functions

These block matching techniques for motion estimation obtain the motion vector by minimizing a cost function. The following cost functions have been proposed in the literature:

(a) The Mean-Absolute Difference (MAD), defined as:

$$MAD(dx, dy) = \frac{1}{mn} \sum_{i=-n/2}^{n/2} \sum_{j=-m/2}^{m/2} |F(i,j) - G(i+dx, j+dy)| \qquad (7.1)$$

where:

F(i, j) - represents a $(m \times n)$ macroblock from the current frame,

G(i, j) - represents the same macroblock from a reference frame (past or future),

(dx, dy) - a vector representing the search location.

The search space is specified by: $dx = \{-p, +p\}$ and $dy = \{-p, +p\}$.

For a typical MPEG system, m=n=16 and p=6, the MAD function becomes:

$$MAD(dx, dy) = \frac{1}{256} \sum_{i=-8}^{8} \sum_{j=-8}^{8} |F(i,j) - G(i+dx, j+dy)| \qquad (7.2)$$

and

$$dx = \{-6, 6\}, dy = \{-6, 6\}$$

(b) The Mean-Squared Difference (MSD) cost function is defined as:

$$MSD(dx, dy) = \frac{1}{mn} \sum_{i=-n/2}^{n/2} \sum_{j=-m/2}^{m/2} [F(i,j) - G(i+dx, j+dy)]^2 \qquad (7.3)$$

(c) The Cross-Correlation Function (CCF) is defined as:

$$CCF(dx, dy) = \frac{\sum_i \sum_j F(i,j)G(i+dx, j+dy)}{\left(\sum_i \sum_j F^2(i,j)\right)^{\frac{1}{2}} \left(\sum_i \sum_j G^2(i+dx, j+dy)\right)^{\frac{1}{2}}} \qquad (7.4)$$

The mean absolute difference (MAD) cost function is considered as a good candidate for video applications, because it is easy to implement it in hardware. The other two cost functions, MSD and CCF, can be more efficient, however are too complex for hardware implementations.

To reduce the computational complexity of MAD, MSD, and CCF cost functions, Ghavani and Mills have proposed a simple block matching criterion, called Pixel Difference Classification (PDC) [GM90]. The PDC criterion is defined as:

$$PDC(dx, dy) = \sum_i \sum_j T(dx, dy, i, j) \qquad (7.5)$$

for $(dx, dy) = \{-p, p\}$.

$T(dx, dy, i, j)$ is the binary representation of the pixel difference defined as:

$$T(dx, dy, i, j) = \begin{cases} 1, & \text{if } |F(i,j) - G(i + dx, j + dy)| \leq t; \\ 0, & \text{otherwise.} \end{cases} \qquad (7.6)$$

where t is a pre-defined threshold value.

In this way, each pixel in a macroblock is classified as either a matching pixel (T=1), or a mismatching pixel (T=0). The block that maximizes the PDC function is selected as the best matched block.

The exhaustive search algorithm

The exhaustive search algorithm is the simplest but computationally intensive search method, which evaluates the cost function at every location in the search area. If MSD cost function is used for estimating the motion vector, it would be necessary to evaluate $(2p + 1)^2$ MSE functions. For p=6, it gives 169 iterations for each macroblock.

The three-step search algorithm

The three-step search algorithm, proposed in [K+81] and implemented in [L+94], first calculates the cost function at the center and eight surrounding locations in the search area. The location that produces the smallest cost function (typically MSD function is used) becomes the center location for the next step, and the search range is reduced by half.

For illustration, a three-step motion vector estimation algorithm for p=6 is shown in Figure 7.4.

Step 1

In the first step, nine values for the cost function MAD (for simplification purposes denoted as M) are calculated: $M_1 = M(0,0)$, $M_2 = M(3,0)$, $M_3 = (3,3)$, $M_4 = M(0,3)$, $M_5 = M(-3,3)$, $M_6 = M(-3,0)$, $M_7 = M(-3,-3)$, $M_8 = (0,-3)$, $M_9 = M(3,-3)$, as illustrated in Figure 7.4. Assuming that M_3 gives the smallest cost function, it becomes the center location for the next step.

Step 2

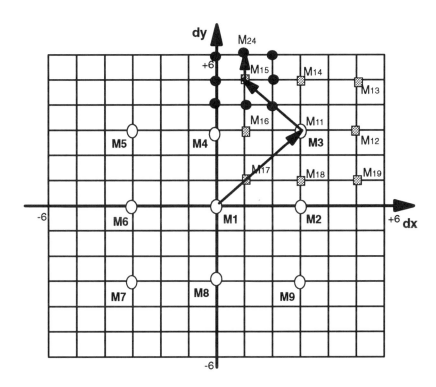

Figure 7.4 The three-step motion vector estimation algorithm - an example.

Nine new cost functions are calculated, for M_3 and surrounding 8 locations, using a smaller step equal to 2. These nine points are denoted in Figure 7.4 as $M_{11}, M_{12}.M_{13}, ...M_{19}$.

Step 3

In the last step, the location with the smallest cost function is selected as a new center location (in the example in Figure 7.4 this is M_{15}), and 9 new cost functions are calculated surrounding this location: $M_{21}, M_{22}, M_{23}, ...M_{29}$. The smallest value is the final estimate of the motion vector. In the example in Figure 7.4 it is M_{24}, which gives the motion vector {dx, dy} equal to {1, 6}.

Note that the total number of computations of the cost function is: $9 \times 3 - 2 =$ 25, which is much better than 169 in the exhaustive search algorithm.

The 2-D logarithmic search algorithm

This algorithm, proposed in [JJ81], uses the MSD cost function and performs a logarithmic 2-D search along a virtual direction of minimum distortion (DMD) on the data within the search area. The modified version of the algorithm described in [SR85], uses the MAD cost function, and can be described using the following steps, as illustrated in Figure 7.5.

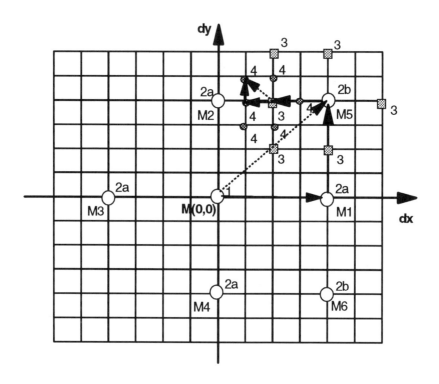

Figure 7.5   The modified 2D logarithmic search algorithm - an example.

Step 1

The MAD function is calculated for $dx = dy = 0, M(0,0)$, and compared to the threshold (lets say the value is 4 out of 255): $M(0,0) < T$. If this is satisfied, the tested block is unchanged and the search is complete.

Step 2a

The next four cost functions are calculated, $M_1(4,0), M_2(0,4), M_3(-4,0)$, and $M_4(0,-4)$, and their minimum is found and compared to $M(0,0)$:

$$M' = min(M_1, M_2, M_3, M_4) < M(0,0).$$

If the minimum $M' > M(0,0)$ go to step 3, otherwise this value is compared against the threshold T. If $M' < T$, the value M' is the minimum and the search ends. Otherwise, the algorithm continues with step 2b.

Step2b

Assuming that in the previous step 2a, the minimum $M' = M_1(4,0)$, then the next two surrounding positions are calculated: $M_5(4,4)$ and $M_6(4,-4)$, as indicated in Figure 7.5. The tests for minimum and threshold are performed again, and if the minimum is found, the procedure is complete. Otherwise, step 3 continues.

Step 3

Assuming that the new minimum location is $M_5(4,4)$, a similar search procedure (steps 2a and 2b) is continued, except the step is divided by 2. In Figure 7.5, the new minimum becomes $M(2,4)$.

Step 4

The step is further reduced by 2, and the final search (steps 2a and 2b) is performed. The minimum (dx, dy) is found; in Figure 7.5 it is (1,5).

For p=6, this algorithm requires maximum 19 cost function calculations, as shown in Figure 7.5.

The conjugate direction search algorithm

This algorithm for motion vector estimation, proposed in [SR85], is an adaptation of the traditional iterative conjugate direction search method. This method can be implemented as one-at-a-time search method, as illustrated in Figure 7.6.

In Figure 7.6, direction of search is parallel to one of coordinate axes, and each variable is adjusted while the other is fixed. This method has been adapted for

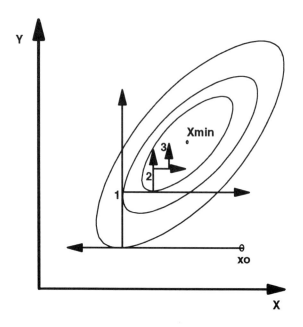

Figure 7.6   A principle of the conjugate direction search algorithm (one-at-a-time search).

motion vector estimation [SR85], as illustrated in Figure 7.7. The algorithm consists of the following three steps:

Step 1

Values of the cost function MAD in the dx direction are calculated, until the minimum is found. The calculation is as follows: (a) M(0,0), M(1,0), and M(-1,0); (b) If M(1,0) is the minimum, M(2,0) is computed and evaluated, and so on. This step is complete when a minimum in the dx direction is found (in Figure 7.7 the minimum is M(2,0).

Step 2

The search now continues in the dy direction by calculating cost functions M(2,-1) and M(2,1). A minimum in the dy direction is then found at M(2,2) in Figure 7.7.

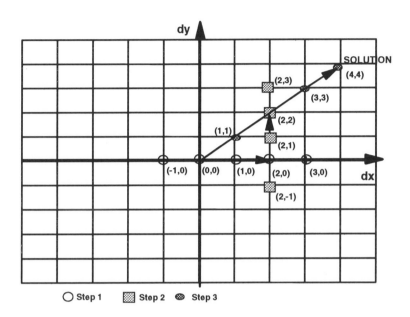

Figure 7.7 The conjugate direction search method for motion vector estimation - an example.

Step 3

The direction of search is now the vector connecting the starting point $(0,0)$ and the obtained minimum $(2,2)$. The following cost functions are calculated and evaluated next: $M(1,1)$ and $M(3,3)$, and so on, until a minimum in this direction is found. In the example in Figure 7.7, the minimum is $M(4,4)$, and the obtained motion vector is $dx = 4$ and $dy = 4$.

It may happen that the dx and dy vectors, obtained in steps 2 and 3, do not constitute a square as given in Figure 7.7. In that case, the nearest grid points on the direction joining $(0,0)$ and the obtained minimum point are selected [SR85].

The parallel hierarchical 1-D search algorithm (PHODS)

The PHODS algorithm, proposed in [C+91], reduces the number of blocks to be searched. In this algorithm, the 2D motion vector for each block is represented as two 1D motion vectors, one in horizontal and one in vertical direction. Both vectors are searched in parallel. The algorithm uses MAD or MSD cost function, and can be described using the following steps, as illustrated in Figure 7.8.

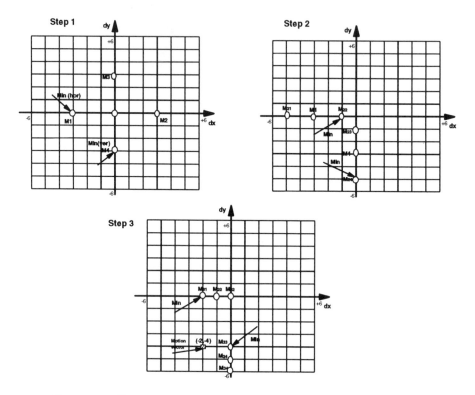

Figure 7.8   The parallel hierarchical 1D search algorithm - an example.

Step 1

The cost function is calculated in the search center (M_0), and in two additional positions in both directions (M_1 and M_2 in horizontal direction, and M_3 and M_4 in vertical direction). The positions with the smallest cost functions in both directions become the search centers in the next step, and the distance between the candidate blocks is reduced. In the example in Figure 7.8, M_1 gives the minimum in horizontal direction, while M_4 in vertical direction. They are new search centers for the next step.

Step 2

The cost function is further calculated in M_1, M_{21}, and M_{22} horizontal locations, and M_4, M_{23} and M_{24} vertical locations. New minimum locations are found as M_{22} in horizontal, and M_{24} in vertical direction.

Step 3

In the last step, the search step is further reduced to 1, and the cost function is calculated in horizontal locations M_{22}, M_{31}, and M_{32}, and vertical locations M_{24}, M_{33}, and M_{34}. Assuming that M_{31} is the minimum in horizontal direction, and M_{33} in vertical direction, the estimated motion vector becomes $dx = -2$ and $dy = -4$.

The modified pixel-difference classification, layered structure algorithm (MPDC-LSA)

This algorithm, proposed in [C+94], is a block-matching algorithm based on a modified PDC function, which gives low computational complexity. The algorithm is based on the MPDC criterion, where a block is classified as a matching block if all its pixels are matching pixels with respect to the defined threshold. The MPDC criterion is defined as:

$$MPDC(dx, dy) = \begin{cases} 1, & \text{if } PDC(dx, dy) = n^2; \\ 0, & \text{otherwise.} \end{cases} \qquad (7.7)$$

The search procedure and experimental results are presented in [C+94].

Using a block-matching motion estimation technique, the best motion vector(s) is found, which specifies the space distance between the actual and the reference microblocks. The macroblock in the current frame is then predicted based on a microblock in a previous frame (forward prediction), a microblock in a future frame (backward prediction), or using interpolation between microblocks in a previous and a future frame. A microblock in the current frame F(i,j) is predicted using the following expression:

$$F_p(i, j) = G(i + dx, j + dy) \qquad (7.8)$$

for $(i, j) = \{-8, 8\}$.

$F_p(i,j)$ is the predicted current macroblock, $G(i,j)$ is the same macroblock in a previous/future frame, and (dx, dy) is the estimated motion vector.

For interpolated frames, a macroblock in the current frame F(i,j) is predicted using the following formula:

$$F_p(i,j) = \frac{1}{2}[G_1(i+dx_1, j+dy_1) + G_2(i+dx_2, j+dy_2)] \qquad (7.9)$$

where $(i,j) = \{-8, 8\}$.

$G_1(i,j)$ is the same microblock in a previous frame, (dx_1, dy_1) is the corresponding motion vector, $G_2(i,j)$ is the same microblock in a future frame, and (dx_2, dy_2) is its corresponding motion vector.

The difference between predicted and actual macroblocks, called the error terms $E(i,j)$, is then calculated using the following expression:

$$E(i,j) = F(i,j) - F_p(i,j) \qquad (7.10)$$

for $(i,j) = \{-8, 8\}$.

Block diagram of a MPEG-1 encoder, which includes motion predictor and motion estimation, is shown in Figure 7.9, while a typical MPEG-1 decoder is given in Figure 7.10.

I frames are created similarly to JPEG encoded pictures, while **P** and **B** frames are encoded in terms of previous and future frames. The motion vector is estimated, and the difference between the predicted and actual blocks (error terms) are calculated. The error terms are then DCT encoded and finally the entropy encoder is used to produce the compact code.

7.2   AUDIO ENCODER AND DECODER

The MPEG standard also covers audio compression. MPEG uses the same sampling frequencies as compact disc digital audio (CD-DA) and digital audio tape (DAT) . Besides these two frequencies, 44.1 KHz and 48 KHz, 32 KHz

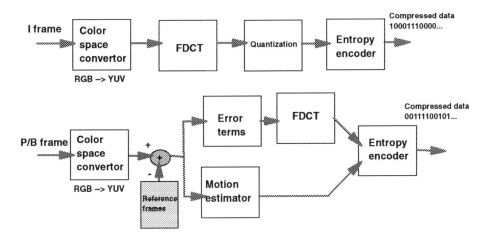

Figure 7.9   The block diagram of a typical MPEG-1 encoder.

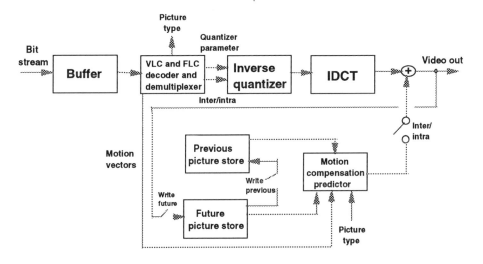

Figure 7.10   The block diagram of a typical MPEG-1 decoder.

is also supported, all at 16 bits. The audio data on a compact disc, with 2 channels of audio samples at 44.1 KHz with 16 bits/sample, requires a data rate of about 1.4 Mbits/s [Pen93]. Therefore, there is a need to compress audio data as well.

Existing audio compression techniques include μ-law and Adaptive Differential Pulse Code Modulation (ADPCM) , which are both of low complexity, low compression ratios, and offer medium audio quality. The MPEG audio compression algorithm is of high complexity, but offers high compression ratios and high audio quality. It can achieve compression ratios ranging from 5:1 to 10:1.

The MPEG audio compression algorithm comprises of the following three operations:

- The audio signal is first transformed into the frequency domain, and the obtained spectrum is divided into 32 non-interleaved subbands.

- For each subband, the amplitude of the audio signal is calculated, as well as the noise level is determined by using a "psychoacoustic model". The psychoacoustic model is the key component of the MPEG audio encoder and its function is to analyze the input audio signal and determine where in the spectrum the quantization noise should be masked.

- Finally, each subband is quantized according to the audibility of quantization noise within that band.

The MPEG audio encoder and decoder are shown in Figure 7.11 [Pen93, Ste94].

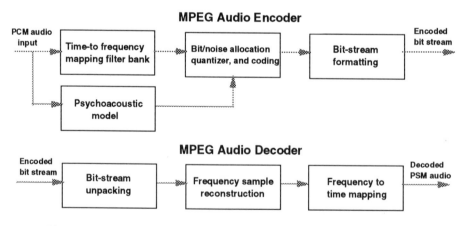

Figure 7.11   Block diagrams of MPEG audio encoder and decoder.

The input audio stream simultaneously passes through a filter bank and a psychoacoustic model. The filter bank divides the input into multiple subbands, while the psychoacoustic model determines the signal-to-mask ratio of each subband. The bit or noise allocation block uses the signal-to-mask ratios to determine the number of bits for the quantization of the subband signals with the goal to minimize the audibility of the quantization noise. The last block performs entropy (Huffman) encoding and formatting the data. The decoder performs entropy (Huffman) decoding, then reconstructs the quantized subband values, and transforms subband values into a time-domain audio signal.

The MPEG audio standard specifies three layers for compression: layer 1 represents the most basic algorithm and provides the maximum rate of 448 Kbits/sec, layers 2 and 3 are enhancements to layer 1 and offer 384 Kbits/sec and 320 Kbits/sec, respectively. Each successive layer improves the compression performance, but at the cost of greater encoder and decoder complexity.

A detailed description of audio compression principles and techniques can be found in [Pen93].

7.3   MPEG DATA STREAM

The MPEG standard specifies a syntax for the interleaved audio and video data streams. An audio data stream consists of frames, which are divided into audio access units. Audio access units consist of slots, which can be either four bits at the lowest complexity layer (layer 1), or one byte at layers 2 and 3. A frame always consists of a fixed number of samples. Audio access unit specifies the smallest audio sequence of compressed data that can be independently decoded. The playing times of the audio access units of one frame are 8 ms at 48 KHz, 8.7 ms at 44.1 KHz, and 12 ms at 32 KHz [Ste94]. A video data stream consists of six layers, as shown in Table 7.2.

At the beginning of *the sequence layer* there are two entries: the constant bit rate of a sequence and the storage capacity that is needed for decoding. These parameters define the data buffering requirements. A sequence is divided into a series of *GOPs*. Each GOP layer has at least one **I** frame as the first frame in GOP, so random access and fast search are enabled. GOPs can be of arbitrary structure (**I**, **P** and **B** frames) and length. The GOP layer is the basic unit for editing an MPEG video stream.

Syntax layer	Functionality
Sequence layer	Context unit
Group of pictures layer	Random access unit: video coding
Picture layer	Primary coding unit
Slice layer	Resynchronization unit
Macroblock layer	Motion compensation unit
Block layer	DCT unit

Table 7.2   Layers of MPEG video stream syntax.

The *picture layer* contains a whole picture (or a frame). This information consists of the type of the frame (**I**, **P**, or **B**) and the position of the frame in display order.

The bits corresponding to the DCT coefficients and the motion vectors are contained in the next three layers: *slice, macroblock, and block layers*. The block is a (8x8) DCT unit, the macroblock is a (16x16) motion compensation unit, and the slice is a string of macroblocks of arbitrary length. The slice layer is intended to be used for resynchronization during a frame decoding when bit errors occur.

7.4   MPEG EXPERIMENTAL RESULTS

In this section we present results obtained using two MPEG-1 software decoders, Berkeley and Hewlett-Packard decoders.

Berkeley Software Decoder

The Berkeley research group has developed the MPEG-1 video software decoder and made it public domain available for research communities [PSR93]. The MPEG-1 decoder is implemented in C using X-Windows system. It is comprised of 12,000 lines of code.

The software video decoder can decode and play 160×120 MPEG-1 video streams in real time at 30 frames/sec on current generation of powerful RISC workstations. It can also decode and play 320×240 video streams at about 16 frames/sec, which is a factor of two of real time. The code was also ported at a proprietary video processor at Philips research, where it is capable of decoding and playing in real-time 352 × 288 video frames.

The decoder has been ported to over 25 platforms and has been distributed to over 10,000 sites over the Internet. We ported the decoder in the Multimedia Laboratory at Florida Atlantic University, and analyzed its performance. The results are presented in this section. First, we analyzed the execution times of different blocks within the MPEG decoder. The obtained results are shown in Table 7.3 [PSR93]. Interestingly enough, reconstruction and dithering require more than 50% of total time due to memory intensive operations. On the other hand, IDCT requires only less than 15% of the total time. The conclusion is that the decoder can be speed up significantly by developing techniques to reduce memory traffic in reconstruction and dithering.

FUNCTION	TIME [%]
PARSING	17.4%
IDCT	14.2%
RECONSTRUCTION	31.5%
DITHERING	24.3%
MISC. ARITHMETIC	9.9%
MISC.	2.7%

Table 7.3   Berkeley MPEG-1 software decoder - analysis of execution times.

In the following experiment, we analyzed the MPEG-1 decoder in decoding five motion-videos which are of different frame sizes and compressed with various compositions of **I**, **P**, and **B** frames. The obtained results are shown in Table 7.4. Bit streams for the first three examples were created by the Xing software encoder, that only produces **I** frames (MJPEG) . The last two streams were produced with a true MPEG encoder.

Note that in the case of a true MPEG decoder (videos 'Flower' and 'Bicycle'), compression ratios for **B** frames were much higher than for **P** frames and **I** frames.

VIDEO	Video Characteristics		Encoder Characteristics					Decoder Performance	
	Frame size	Total No of frames	I/P/B composition	Number of MBs/frame	Compression ratio	Comp. ratio for I/P/B	Compressed bits/pixel	Decoding time [s]	Average frame/sec
Grand	160x120	194	194/0/0	80	40.3	40.3/-/-	0.60	16.7	11.6
Mjackson	160x120	557	557/0/0	80	47.4	47.4/-/-	0.51	40.4	13.8
Rom	160x120	73	73/0/0	80	17.9	17.9/-/-	1.34	4.2	17.4
Flower	352x240	148	10/40/98	330	52.9	14.6/31.5/ 114.9	0.45	58.3	2.5
Bicycle	352x240	148	10/40/98	330	47.4	24.9/36.2/ 76.3	0.45	48.0	3.1

Table 7.4   Performance analysis of the Berkeley software MPEG-1 decoder.

In the final experiment, we measured the performance of the Berkley MPEG-1 software decoder running on different computer platforms, for two MPEG video streams. The results are presented in Table 7.5.

VIDEO	Compression Ratio	PERFORMANCE [frames/sec]			
		DEC AXP 3000/500 @ 150 MHz	HP 9000/750 @ 66 MHz	SUN SPARC 10/30 @ 36 MHz	Intel 486 DX/2 @ 66 MHz
Canyon 144x112	49	43.1	74.7	13.4	38.1
Flower 320x240	50	8.9	15.4	3.3	8.2

Table 7.5   Berkeley software MPEG-1 decoder - comparison of different computer platforms.

From Table 7.5, it can be concluded that three out of four tested machines (except Sun SPARC) can decode 144 × 112 video sequences in real-time (30 frames per second or more). However, for video sequences 320 × 240 the best performance is within factor of two of real-time performance (HP9000/750 has achieved 15.4 frames/sec).

Hewlett-Packard MPEG-1 Software Decoder

Hewlett-Packard has recently implemented a MPEG-1 software decoder on its PA-RISC processor, which is capable of decompressing MPEG-1 videos at real-time rates of 30 frames/sec [Lee95]. The MPEG-1 software video player is written in C and runs on an enhanced PA-RISC general purpose microprocessor.

The PA-RISC architectural enhancements, implemented in a PA-7100LC processor, which is the central processor of the HP712 desktop computer, includes the following features:

- New multimedia instructions, implemented in PA-7100LC, allow simple arithmetic operations (on 16-bit data) to be executed in parallel,

- The processor has two Arithmetic and Logic Units (ALUs), which allow a parallelism of four 16-bit operations per cycle,

- The graphics subsystem has been enhanced by implementing the color conversation step together with the color recovery step. Color conversation converts between the YcbCr and the RGB color formats, while color recovery allows 24-bit RGB color that had been "color compressed" into 8 bits to be converted back to 24-bit color before being displayed (dithering operation). This enhancement reduces the display operation in the MPEG decoding.

- In the PA-7100LC processor, the memory controller and the I/O controller have been integrated, which significantly reduces the overhead in the memory to frame buffer bandwidth.

The reported results in [Lee95], shown in Figure 7.12, proved that the described enhancements, implemented in the HP712 desktop computer, provide much higher performance over the older high-end HP735 workstation running at 99 MHz and HP715 running at 50 MHz.

Figure 7.13 compares results, reported in the literature [Lee95], obtained by running MPEG-1 software decoders on various processors. These results are

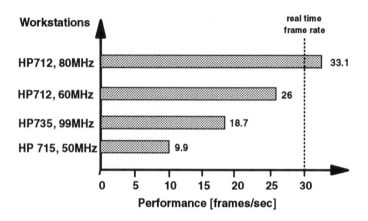

Figure 7.12   Performance of the Hewlett-Packard MPEG-1 software decoder [Lee95].

not verified. Note that besides the HP712 processor at 80 MHz, the DEC Alpha processor at 175 MHz was also able to run the video at a real time speed of 30 frames/sec.

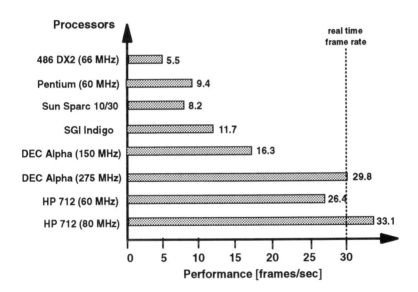

Figure 7.13 Comparison of various processors running MPEG-1 software decoder [Lee95]. (Results are not verified).

8

OTHER MULTIMEDIA
COMPRESSION TECHNIQUES

In this chapter we briefly introduce four other multimedia compression techniques:

- Digital Video Interactive (DVI) technology,

- Fractal image compression

- Subband image and video coding, and

- Wavelet-based compression.

DVI is already a mature technology implemented in IBM's multimedia systems, while the other three are promising compression techniques, which are still in the research stage.

8.1 DIGITAL VIDEO INTERACTIVE (DVI) TECHNOLOGY

Digital Video Interactive (DVI) technology has been originally developed at David Sarnoff Research Center of RCA in Princeton, and later has been taken over by Intel and IBM. It is presently used as a codec technology for the IBM's ActionMedia audio/video capture and playback boards for PS/2-based personal computers. The DVI system has been implemented using two specially-designed processors: the 82750PB pixel processor (PB), and the 82750DB display processor (DB), as shown in Figure 8.1 [Rip89, H+91].

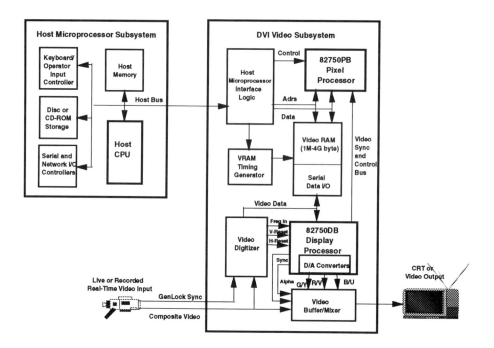

Figure 8.1 The video subsystem based on DVI technology.

The pixel processor performs image compression and decompression, generates fast graphics and special effects, and acts as the arbiter for memory and device transactions. The display processor performs various real-time display functions, such as data format transformations, color translation, interpolation of pixel values, and provides outputs to the display devices [H+91].

DVI technology performs its operations in an internal YUV format, in which a pure white pixel consists of 30% red, 59% green, and 11% blue components. A RGB signal is then transformed into the YUV signal using the expressions, given in Section 4.3. This color space conversion is accomplished by performing matrix multiplication on the YUV pixels.

The pixel processor 82750PB is a microcoded machine with a 16-bit specialized ALU, a 16-bit barrel shifter and two specialized units optimized for image processing tasks. The first unit, the pixel interpolator, calculates fractional spatial movement of pixels, scales groups of pixels, and performs various filtering operations. The second unit, the statistical decoder, decompresses video data in real time.

DVI Compression Algorithm

The DVI system uses a combination of several compression techniques, but the key algorithm for compression of motion video is the Production Level Video (PLV). The PLV algorithm is based on motion-compensated prediction and vector quantization [H+91]. In this algorithm, most frames are encoded in reference to the previous frame in the video sequence, and the motion vectors are calculated, similar to the MPEG and H.261 techniques.

Furthermore, the DVI technique achieves a high compression ratio by specifying the region in the previous frame in fractional pixel coordinates. This means that the DVI decoder must first interpolate between pixel values in the previous frame to derive the starting region in the current frame. This function is performed by the pixel interpolator, implemented in the pixel processor (PB), and as the result the motion-compensated array is derived.

In the second step, the decoder constructs the array of signed difference pixels and adds them to the motion-compensated array to produce the final reconstructed region. In the PLV algorithm, the difference array between pixels in a region can be constructed using one of several algorithms. In 2×1 vector quantization algorithm, described in [H+91], each horizontally adjacent pair of pixels in the difference array is quantized to one of a finite set of number pairs (vectors). Each vector in the set is identified by a unique integer which is then statistically encoded. The decoder simply decodes a Huffman code and does a lookup into the current table of vectors. The pair of numbers retrieved from the table is then written to the difference array and added to the pixel values from the motion-compensated region of the previous frame. Figure 8.2 illustrates the decoding process.

In the number example in Figure 8.2, the decoder first decodes the Huffman code $110 = 6$, and then on the basis of this code obtains the pair of numbers (3,-4) from the vector lookup table. These numbers are then written in the difference array, and added to the corresponding pixel values from the motion-compensated region of the previous frame (42,93). The pixels in the current frame are then constructed as (45,89).

The 82750PB pixel processor is capable of performing the described decoding process using only four instructions for every two pixels [H+91]. Therefore, a

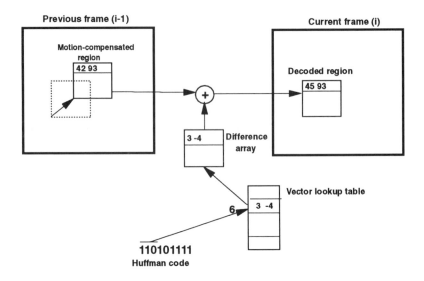

Figure 8.2   Interframe motion-compensated vector quantization implemented in DVI.

25 MHz pixel processor can process about 200,000 pixels in 1/60 of a second. Assuming that the decompression of an entire frame includes additional steps, the total execution speed is about 70,000 pixels every 1/60 second, which is sufficient for a good quality full-screen video.

8.2   FRACTAL IMAGE COMPRESSION

The JPEG image compression technique, described in Chapter 5, is very effective at low compression ratios of about 25:1. The most serious problem of JPEG compressed images is that they are resolution dependent. If the decompressed image is displayed at a higher resolution than the original, it will result in the blockiness due to pixel replication. Since graphics cards and printers are increasing in resolution constantly, the resolution dependence of JPEG images results in needs to rescan and recompress them constantly in order to take advantage of the latest graphics and printers' technologies.

A fractal is defined as an infinitely magnifiable picture that can be produced by a small set of instructions and data. The fractal compression is based on breaking an image into pieces (fractals), and then identifying fractals which are similar or identical [BH93, Ans93, Fis94]. For example, the image of 'Lena', shown in Figure 8.3, shows various regions which are self-similar (similar to other parts of the image). The smaller block is, it is easier to find other portions of the image that are similar to each other. The image is divided into domain and range blocks. For each domain block, a range block is found, that after transformation matches the domain block as closely as possible.

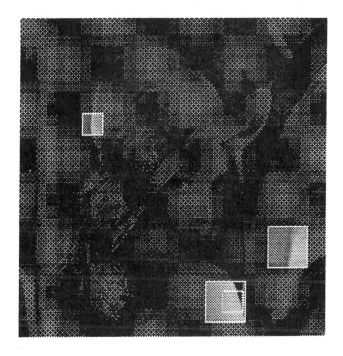

Figure 8.3 The image 'Lena' partitioned into regions shows self-similarity portions.

Figure 8.4 illustrates several ways to divide image 'Lena' into domain and range blocks based on rectangles and triangles partitioning, proposed in [Fis94, MF94b].

Fractal compression is based on affine transformations, which is a mathematical function consisting of a combination of rotation, scaling, skew, and translation in n-dimensional space. An example of an affine transformation in two dimensions is given as:

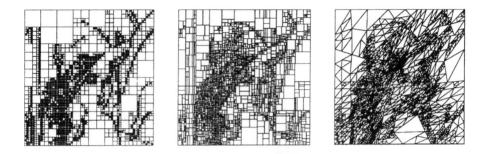

Figure 8.4 Dividing the image 'Lena' into domain and range blocks - rectangles and triangles partitioning.

$$W(x, y) = (ax + by + e, cx + dy + f) \qquad (8.1)$$

or in matrix form:

$$W \begin{bmatrix} x \\ y \end{bmatrix} = \begin{bmatrix} a & b \\ c & d \end{bmatrix} \begin{bmatrix} x \\ y \end{bmatrix} + \begin{bmatrix} e \\ f \end{bmatrix} = \begin{bmatrix} ax + by + e \\ cx + dy + f \end{bmatrix} \qquad (8.2)$$

The matrix

$$\begin{bmatrix} a\ b \\ c\ d \end{bmatrix}$$

determines the rotation, skew, and scaling, and

$$\begin{bmatrix} e \\ f \end{bmatrix}$$

determines the translation.

This transformation moves the point (0, 0) to (e, f), the point (1, 0) to (a + e, c + f), the point (0, 1) to (b + e, d + f), and the point (1, 1) to (a + b +

e, c + d + f). The values a, b, c, d, e, and f are the affine coefficients for this transformation.

Fractal compression consists of finding the right set of affine transformations. The compression algorithm for fractal compression is shown in Figure 8.5 [Ans93].

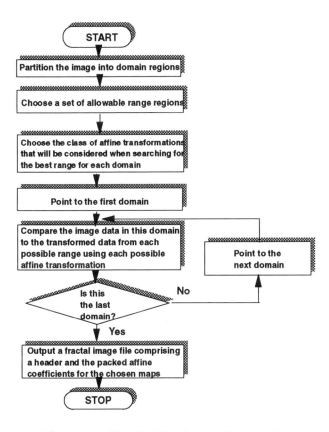

Figure 8.5 The algorithm for fractal compression.

The image is first partitioned into non-overlapping domain regions as discussed earlier. This set of domain regions must cover the entire image, but they can be any size or shape. In the next step, a collection of range regions are defined, which are larger than domain regions, can overlap and need to cover the entire image.

In the next step, the algorithm chooses the class of affine transformations, which when applying on the range regions, closely match the domain region. Each 3-D affine transformation is defined by its affine coefficients.

Finally, a Fractal Image Format (FIF) is written, which consists of a header with information on the specific choice of domain regions, followed by the packed list of affine coefficients selected for each domain region.

This compression process generates a file that is independent on resolution of the original image, and identified parameters specify an *equation for the image*. For example, a straight line can be represented by the equation $y = ax + b$. By knowing values for coefficients a and b, the line can be drawn at any resolution. Similarly, for given affine coefficients of an image, the decompression algorithm can create a fractal replication of the original image at any resolution. The algorithm for fractal decompression is shown in Figure 8.6 [Ans93].

The decompression process is an iterative process, where each domain region is replaced with the transformed data from the appropriate range region using the affine coefficients stored for this domain.

The fractal compression/decompression process is asymmetric, much more computation is required for compression than for decompression. In addition, compression ratios can be improved by increasing compression time, with no increase in decompression time or decrease in image quality.

A number of experiments comparing fractal and JPEG compressions have been performed [MF94b, Ans93]. In the case of an 800×600 color image, which requires 1.44 MB of disk space, the JPEG software compression and decompression took 41 seconds each on a 386/33 MHz system [Ans93]. The fractal image compression was done in 8 minutes, but the decompression took only 7 seconds.

In experimental results, reported in [MF94b] on several 320×200 grayscale images, fractal compression ratio was in average 15, while JPEG compression ratios varied between 3 and 20, depending on selected image quality.

In summary, in contrast to JPEG, fractal compression offers fast decompression, resolution independence, and high compression ratios. It can be useful for asymmetric multimedia applications where quick access to high-quality images is important.

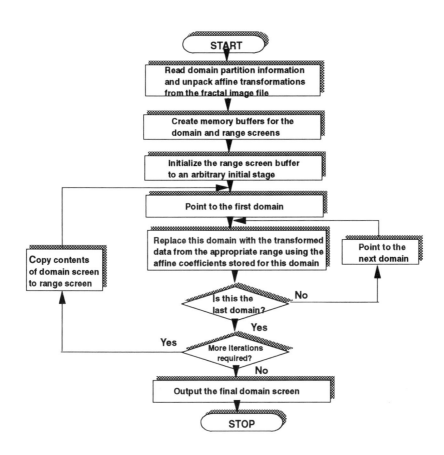

Figure 8.6 The algorithm for fractal decompression.

8.3 SUBBAND IMAGE AND VIDEO CODING

In the subband coding method for image and video compression, the block transform technique (such as the DCT) is replaced with the subband transform technique. The concept of subband coding is based on splitting the input signals into smaller frequency bands, these subbands are then encoded, transmitted or stored, and later reconstructed using the decoded subbands [Woo91]. The commonly used transform method for image and video subband coding based on a 2-D Quadrature Mirror Filtering (QMF). The 2-D QMF is performed by applying 1-D QMF along the rows and then along the columns, as shown in Figure 8.7.

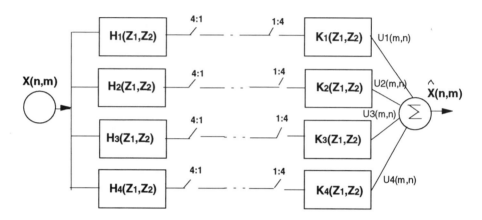

Figure 8.7   Two-dimensional QMF bank system.

In the 2-dimensional QMF system from Figure 8.6, $H_i(z_1, z_2)$ and $K_1(z_1, z_2)$ are z transforms of two filters $h_i(n, m)$ and $k_i(n, m)$, for $i = 1, 2, 3, 4$. The following conditions must be satisfied in order to realize an alias-free reconstruction of the input signal $X(n, m)$ at the receiver:

$$H_2(z_1, z_2) = -K_2(z_1, z_2) = H_2(-z_1, z_2)$$
$$H_3(z_1, z_2) = -K_3(z_1, z_2) = H_1(z_1, -z_2)$$
$$H_4(z_1, z_2) = K_4(z_1, z_2) = H_1(-z_1, -z_2)$$
$$H_1(z_1, z_2) = -H_{11}(z_1)H_{12}(z_2) \qquad (8.3)$$

The separability characteristics, specified as:

$$H_1(z_1, z_2) = H_{11}(z_1)H_{12}(z_2) \qquad (8.4)$$

reduces 2-D QMF filtering problem into 1-D filtering, and thus the conventional 1-D QMF techniques can be used. Now, by applying 1-D QMF along the rows and then along the columns, a basic four-band decomposition can be obtained [Vat84], as shown in Figure 8.8a. Figure 8.8b shows the subband reconstruction by using an inversed process, first applying 1-D QMF along the columns, and then along the rows.

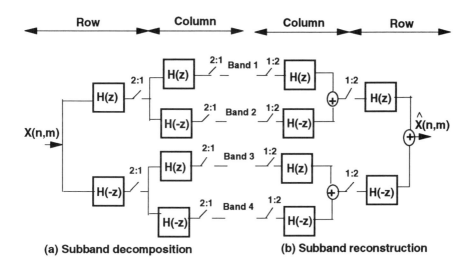

(a) Subband decomposition   (b) Subband reconstruction

Figure 8.8   (a) Subband decomposition, (b) Subband reconstruction.

8.3.1   Subband Coding of Images

The input image is first split into subbands using the subband transform technique described earlier, and then the subbands are encoded using commonly used techniques, such as PCM, DPCM, vector quantization, adaptive predictive coding, or some combinations of these techniques [Woo91, GT88]. For example, in [GT88] 512 × 512 images are split into 7 subbands, and then the subband with the lowest frequencies is encoded using DPCM, while other subbands are encoded using the PCM technique, as illustrated in Figure 8.9.

The selection of coding methods for different subbands can be made based on subband statistics [Woo91, GT88].

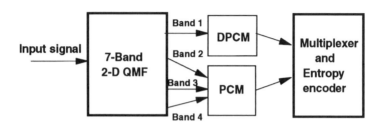

Figure 8.9   Subband encoder of images.

8.3.2   Subband Coding of Video Signals

Several subband compression schemes have been developed for coding of video signals. The system, proposed in [Gha91], represents motion-compensated interframe subband coding system, based on the same principles as px64 or MPEG systems, described in Chapters 6 and 7, respectively. The block diagram of the motion-compensated interframe subband encoder is shown in Figure 8.10.

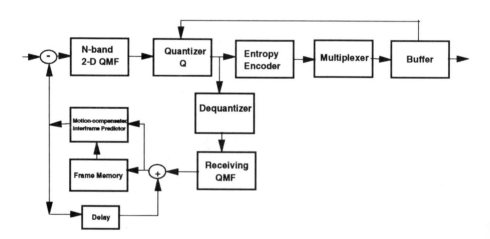

Figure 8.10   The block diagram of the motion-compensated interframe subband encoder.

In this system, the first frame of each sequence is coded using the image subband coding algorithm, described in 8.3.1, while the subsequent frames are coded using motion compensated interframe interpolation, described in Chapter 7. In this model, the signal is decomposed into subbands by QMF and then quantized. The quantized bands are entropy-encoded, multiplexed, and transmitted to the receiver. For the reconstruction of the signal, the coded signal is dequantized, passed through a set of interpolation QMF filters for subband reconstruction, and added to the motion-compensated estimate by passing through an interframe predictor.

Another model for subband encoding of video signals involves decompressing the input video into smaller bands by using 2-dimensional QMF bands. Then, the decompressed bands are passed through DPCM loops where they are subtracted from the motion-compensated estimates [Woo91].

8.3.3   Experimental Results

In this section, we present results obtained in compressing 'Lena' image using the subband coding technique. A detailed analysis is given in [CHF94]. We divided the original image into four subbands: high, mid, low, and base bands, and then the entropy coding is used to encode the individual bands. The entropy encoding used in the experiment is a modified version of the Lempel-Ziv algorithm.

In Table 8.1, the compression ratios for individual bands are calculated, and then the total compression ratio is obtained by dividing the total input pixels by the sum of all compressed pixels from four subbands. In the same table, execution times for the software subband encoder are reported. The decoding times were not reported, but they are close, or in most cases smaller than the encoding times.

In the same table, we reported compression results when the original image 'Lena' is encoded using the entropy encoder only. The obtained compression ratios (2.54 for the entropy encoder versus 16.08 for the subband coder), and for the execution times (17.23 sec for the entropy encoder versus 6.77 sec for the subband coder) show the superiority of the subband coding technique. In both cases the quality of the image was equivalent to the original image. Compared to JPEG and MPEG compression techniques, the subband compression technique offers better compression ratios when a high quality of the image is required (equivalent to the original image). In these cases, a lossless JPEG gives about 1:2 compression ratio, while the subband coding can achieve compression ratios of about 1:5.

	Original number of pixels	Compressed number of pixels	Compression Ratio	Software encoder execution time [sec]
Subband Coding (incl. Entropy Encoding)	836,150	52,003	16.08	6,07
Entropy Encoder only	836,150	328,419	2.54	17.23

SUBBAND CODING	Number of pixels after dividing in subbands	Number of pixels after entropy encoding	Software encoder execution time [sec]
High subband	213,558	33,444	4,39
Medium subband	59,958	12,997	1.20
Low subband	59,958	3,832	0.32
Base subband (lowest)	3,764	1,730	0.16
TOTAL	290,644	52,003	6.07

Table 8.1   Subband coding of image 'Lena'.

The recent focus of research has been to use the subband coding for HDTV applications [Woo91, Gha91, CHF94]. For such applications, subband coding techniques that exploit temporal redundancy can be attractive.

In addition, compared to other techniques which use DCT-based coding with complex computations, the subband coding simply performs the spatial filtering of the input signal into the subbands. This fact makes the subband coding attractive for data transmission in real time. Reported transmission rates using the motion compensated interframe subband coding, described in 8.3.2, are ranging from 6.4 Kb/s to 1.54 Mb/s [Woo91]. These results are obtained using simulation. The current research goal is to achieve the transmission rate of 10 Mb/s. Very high quality subband-based compressors, which can be used for HDTV, are proposed in [Gle93]. The subband image coder, which is capable to preserve the scalability of the representation, is presented in [TSB95].

8.4   WAVELET-BASED COMPRESSION

In the last several years, the wavelet-based image and video compression has become a very challenging research topic [LK90, HJS94]. Several wavelet-based image compression algorithms have been proposed in the literature, and they use various schemes from simple entropy encoding, vector quantization, adaptive transform, to tree encodings and edge-based coding. The block diagram of all these wavelet-based image compression algorithms can be generalized, as shown in Figure 8.11 [HJS94].

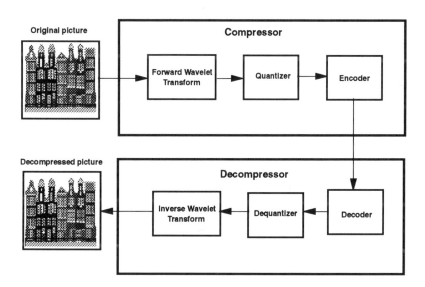

Figure 8.11   The block diagrams of the wavelet-based compressor and decompressor.

The wavelet-based compression consists of three steps: (a) forward wavelet transform to decorrelate the image data, (b) quantization of the transform coefficients, and (c) encoding the quantized coefficients. The decompression consists of inverting compression operations.

Wavelet transform is a type of signal representation which can give the frequency content of the signal at a particular instant of time. In contrast to the wavelet transform, the Fourier representation reveals information about the signal's frequency domain, while the impulse function response describes the time domain behavior of the signal. One of possible implementations of wavelet transform can be using Quadrature Mirror Filters, described in 8.3, and therefore wavelet-based compression can be viewed as a type of subband-based-coding. An example of the wavelet-based image compression using Daubechie's W_6 wavelet is given in [HJS94].

9

IMPLEMENTATIONS OF COMPRESSION ALGORITHMS

In this chapter we describe some commercial implementations of compression algorithms. We also present performance results reported in commercial and research literature on some of these systems.

When implementing a compression/decompression algorithm, the key question is how to partition between hardware and software in order to maximize performance and minimize cost. Most implementations use specialized video processors and programmable Digital Signal Processors (DSPs), however with powerful RISC processors the sole software solutions are becoming feasible.

Figure 9.1 shows the PC-based video compression trends [Ohr93b], where the present, expensive solutions use dedicated video compression chips and DSPs on PC backplanes and motherboards. However, in the near future, more powerful processors will perform real-time compression and decompression in software, thus offering less expensive solutions.

The primary digital signal processing techniques for video compression include image acquisition, temporal prediction, spatial frequency decomposition (such as DCT), quantization, and entropy encoding. At the present state of processor's technology, DSP VLSI chips specially architected to implement video compression algorithms are needed.

Presently, several vendors have developed highly integrated solutions. For example, C-Cube Microsystems and LSI Logic offer both single chips JPEG and MPEG-1 algorithms. Another VLSI approach, taken by Integrated Information Technology (IIT) is based on the Vision Processor (VP), which is a high-speed programmable signal processor that implements both DCT and motion com-

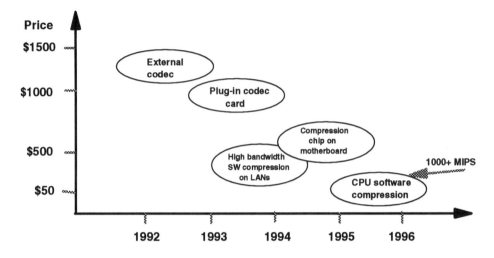

Figure 9.1   PC-based video compression trends.

pensation. In companion with the Vision Controller, the VP can implement several standard algorithms such as MPEG, JPEG, and H.261. Similarly, the AT&Ts AVP-1000 chip set provides multi-algorithm solutions.

There is a class of algorithms that can be implemented today in software with general-purpose processors. They are primarily designed for video decoding allowing playback at real-time frame rates. Typical resolutions of these algorithms are (160 × 120) and (320 × 240) pixels, and the frame rates are about 15 frames/sec. Examples of these software video compressors include Apple Video, Supermac Compact Video (as part of Quick Time), Intel's Indeo, and Microsoft Video, but real-time software MPEG encoders are still years away.

Finally, some advanced processors, such as UltraSPARC, HP-PA-7100LC, and PowerPC620, use enhanced architectural features to provide real-time MPEG decoding and some other functions needed in multimedia applications.

In summary, implementations of compression algorithms can be classified into three categories:

1. Using hardwired approach or single-function processors (the first generation of compression chips) that implement the most demanding operations of algo-

rithms, such as DCT, motion estimation and quantization. Examples include compression chips from C-Cube, LSI Logic, and SGS-Thompson.

2. Multi-standard compression chips (the second generation of compression chips) that, in addition to hardwired engines for computational intensive operations, integrate programmable processors that can be programmed to control the data flow within the chip and support software programming. Examples include compression chips from IIT, Array Microsystems, and C-Cube.

3. Using a software solution and a general purpose processor to maximize flexibility and reduce cost. Examples include Apple Video, Microsoft Video, and Intel Indeo.

9.1 SOFTWARE JPEG SYSTEMS

Three JPEG systems analyzed and evaluated in [Gil91], are presented in this section. These systems are Alice 200 from Telephoto, the Super Still-Frame Compression Card from New Media Graphics, and Optipac 3250 and 5250 compression accelerator boards from Optivision. The results, reported in [Gil91], are presented in Tables 9.1, 9.2 and 9.3.

Image Compression	Size	Compression Ratio	Decompress & Display Time [sec]	File to File Compress Time [sec]	Display to File Compress Time [sec]
Original	409,618	0	3.35		
Perfect (1.5)	175,138	2.34	7.47	15.54	17.36
Best (10)	44,432	9.22	2.31	7.65	6.93
Better (20)	27,484	14.9	1.92	6.99	6.15
Good (32)	19,912	20.57	1.75	6.84	5.93
Worst (199)	7,062	58	1.42	6.10	5.33

Table 9.1   JPEG results for Telephoto Alice 200 [Gil91].

The best performance in terms of speed was achieved with the Optivision 5250, then with the Telephoto Alice 200 and finally with the New Media Super Still-

Image Compression	Size	Compression Ratio	Decompress & Display Time [sec]	File to File Compress Time [sec]
Original	409,618	0	3.35	
JPEG Best (100)	84,272	4.86	15	26
JPEG Better (90)	45,788	8.95	11	30
JPEG Good (75)	33,104	12.37	9	18
JPEG Worst (10)	15,708	26.08	6	19
C-Cube Best	86,284	4.75	3	9
C-Cube Better	44,520	9.20	3	11
C-Cube Good	32,754	12.51	3	9

Table 9.2   JPEG results for New Media Super-Still Frame board [Gil91].

Frame card. For example, compressing from the TARGA display to disk took the Optivision card 1.1 seconds, Telephoto took 5.93, and New Media trailed at 9 seconds.

In another analysis, various JPEG compression programs for DOS and Windows have been tested [Rau92]. A large 8x10 inch color photo was scanned at 266 dpi, resulting in a file of over 14 MB. The image has been compressed and then decompressed for three different levels of image quality. These three levels corresponded to high quality, medium quality, and low quality. The results are presented in Table 9.4.

The best results were obtained using Alice-DOS utility from Telephoto, but in that case H350 accelerator board was used. It compressed a 14 MB file to 143 KB (compression ratio about 10) in a little more than a minute. The best pure software JPEG has been LeadView 1.7 from Lead Technologies, achieving compression in little more than 2 minutes.

Autograph International (AGI) also offers JPEG compression software called EasyTech/Codec. Its interesting feature is that the EasyTech/Codec provides

Image Compression	Size	Compression Ratio	Decompress & Display Time [sec]	File to File Compress Time [sec]	Display to File Compress Time [sec]
Original	409,618	0	3.35		
High Speed	223,646	1.83	3	8	2.03
High Compression	224,020	1.83	10	21	16.75
Best (9)	90,044	4.55	6	7	2.52
Good (5)	17,628	23.24	3	6	1.10
Worst (0)	3,376	121.33	2	7	1.10

Table 9.3   JPEG results for Optivision 5250 [Gil91].

all four modes of JPEG compression: baseline, sequential, progressive, and lossless modes. Compression or decompression of a 512x512 pixel RGB image (about 0.75 MB) is typically performed in 3.5 seconds on a Sun Sparc station including I/O overhead and color conversation between RGB and YUV.

9.2   COMPRESSION CHIPS

In this section we present the design and features of several commercial VLSI compression chips, summarized in Table 9.5. It should be added, that many companies (such as C-Cube, Philips, AT&T, LSI Logic, and SGS Thompson) are currently working on MPEG-2 chips.

9.2.1   AT&T Video-Codec Chip Set

The AT&T has developed two specialized chips, video encoder AVP 4310E and video decoder AVP 422D for full-motion video px64 and MPEG standards [A+93a]. Computer-intensive functions, such as motion estimation and Huffman coding, are implemented in hardware. Key parameters, such as frame rate, delay, bit rate, and resolution, are user programmable. Less stable functions

	Setting	Compres-sed File Size [KB]	Compres-sion Time [min:sec]		Setting	Compres-sed File Size [KB]	Compres-sion Time [min:sec]
AliceDOS	10	560	3:21	LeadView 255	25	409	7:17
	45	195	2:56		115	150	7:21
	80	144	2:52		200	123	7:17
AliceDOS w/H350 accelerator	10	556	1:10	Picture Packer 3.5	1	630	5:45
	45	192	1:04		4	467	5:39
	80	143	1:03		8	410	5:37
ImagePrep 4.0	2	944	6:00	Picture Packer w/ accelerator	1	1,168	3:38
	9	495	5:45		4	467	3:27
	16	262	5:20		8	410	3:28
LeadView 1.7	25	256	2:29	VT Compress 1.1	10	270	4:05
	115	108	2:27		45	149	4:00
	200	97	2:27		80	137	3:55

Table 9.4 Evaluation of JPEG systems [Rau92]. (8x10 inch color photo scanned at 266 dpi; 14 MB file).

are implemented on a programmable RISC processor. The block diagram of the video encoder AVP 4310E is shown in Figure 9.2. The chip was designed as a full-custom device using symbolic layout and compaction, and was built using a 0.9 μm CMOS process. The 120 mm^2 die contains about 1.6 million transistors [A+93a].

The encoder accepts video input in YCbCr format at 30 frames/sec on either its video input bus or the host bus, and outputs compressed data (MPEG or px64) at selectable data rate from 40 Kb/s to 4 Mb/s. The encoder supports video resolutions up to 288×360 and still-image resolutions up to 1024×1024. The encoder uses predictive frame coding based on exhaustive search-motion estimation with a search range of ± 15 pixels.

The video decoder chip AVP-4220D, shown in Figure 9.3, accepts compressed videos (MPEG or px64) at rates up to 4 Mb/s from either serial bus or parallel host bus via the host/serial interface. The decoder outputs the decoded video in raster-scanned order on either its host bus or the dedicated pixel bus. The output is selectable, and can be either 24-bit RGB or 4:2:2 YCrCb format. The decoder requires a 1 MB memory buffer to store previous frames for predictive

COMPANY	COMPRESSION CHIPS
AT&T	AVP-4310E Video encoder AVP-4220D Video decoder for MPEG and H.261
LSI Logic	L64702 JPEG coprocessor L64111 MPEG audio decoder L64112 MPGE video decoder
Texas Instruments	MVP Multimedia Video Processor for MPEG compression
ITT	VP Video Processor for MPEG, JPEG, and H.261

Table 9.5   Compression chips.

and interpolative decoding. The chip was designed using full custom and standard cell technology. It uses 0.9 μm CMOS process and contains 1.2 millions transistors [A+93a].

AT&T has also developed the system controller chip AVP-4120C, a multiple-protocol communications controller for interactive video and videoconferencing. The controller provides a number of functions including multiplexing and demultiplexing px64 or MPEG audio, video, and user data, and audio/video synchronization.

An example system, which uses the AT&T video codec chip set, is a Production Catalysts' video-telephony and multimedia system for PCs [Fri92]. The system, shown in Figure 9.4, comprises of encoder, decoder, and ISDN interface boards. The encoder board consists of the video encoder chip, two DSP 3210 for audio compression, video input subsystem to receive data from a VCR or a live camera, and microphone control. The decoder board contains the video decoder chip, a DSP 3210 chip for audio decompression, and NTSC encoder to output digital video to a VCR. The ISDN board provides digital communications via ISDN networks.

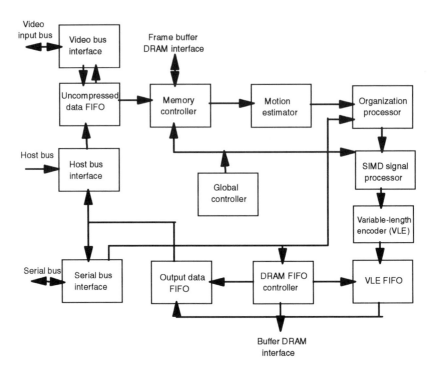

Figure 9.2   The block diagram of the AT&T video encoder AVP-4310E [A+93a].

9.2.2   LSI Logic JPEG and MPEG Chip Sets

LSI Logic has developed several chips and corresponding boards for still-image and video compression and decompression. They are described in this section.

JPEG Coprocessor

LSI Logic's L64702 JPEG coprocessor has been developed for low-cost PC based multimedia applications. It compresses and decompresses a 320x240 image in 20 milliseconds, and therefore it can support real-time video applications (30 frames/sec) as well as still-image applications. A detailed block diagram of the JPEG coprocessor is shown in Figure 9.5 [JV93].

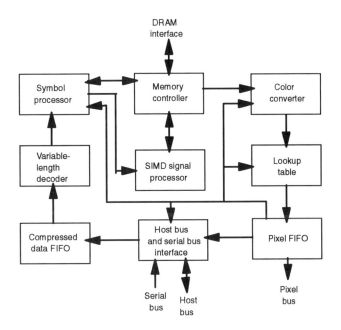

Figure 9.3 Block diagram of the AT&T video decoder chip AVP-4220D [A+93a].

The JPEG processing core unit performs crucial JPEG compression and decompression operations, such as FDCT and IDCT, quantization, DPCM and IDPCM, run length coding, zig-zag conversion, and variable length encoder and decoder. The JPEG processing core unit communicates with two independent interfaces, the video port and the system port.

MPEG Audio and Video Decoders

LSI Logic has also developed two VLSI chips for MPEG audio and video decoding: L64111 MPEG Audio Decoder, and L64112 MPEG Video Decoder targeted for studio-quality compressed digital video broadcast and entertainment applications.

The MPEG Video Decoder supports image resolutions up to 720 × 480 at 30 frames/sec for NTSC system, and 720 × 576 at 25 frames/sec for PAL system. Besides MPEG-1 video decoding, it provides other functions such as support

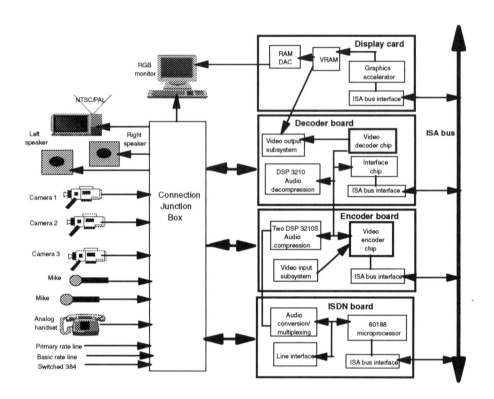

Figure 9.4 Multimedia system developed by Production Catalysts uses AT&T video codec chip set [Fri92].

for channel switching and virtual channels, VCR anticopy schemes, audio/video synchronization, time base correction, built-in error concealment, and other video timing signals.

The MPEG audio decoder implements MPEG-1 bit stream layer I and II audio decoding in the range from 32 to 192 Kb/s mono and 64 to 384 Kb/s stereo. Output sampling is at 32 KHz, 44.1 KHz, or 48 KHz.

These MPEG audio and video decoders support a number of applications such as decoder boxes for cable TV, wireless CATV, business TV, video-on-demand systems, and other entertainment applications. Figure 9.6 shows a block diagram of a typical application based on this MPEG chip set.

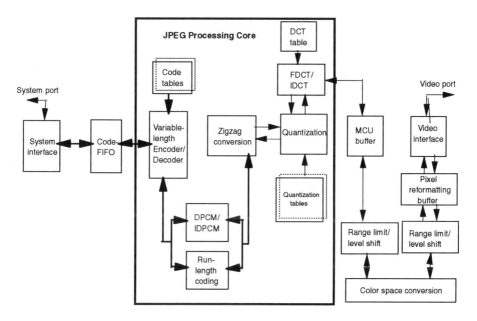

Figure 9.5 Block diagram of the LSI Logic's L64702 JPEG coprocessor. Courtesy of LSI Logic Inc.

9.2.3 Texas Instruments Multimedia Video Processor

Texas Instruments has developed a single chip multiprocessing device, called Multimedia Video Processor (MVP), optimized for video and image processing [L+94, GGV92]. The MVP combines a RISC and four DSP processors in a crossbar-based shared-memory parallel architecture, as shown in Figure 9.7.

The processor and the memory modules are fully interconnected through the crossbar which can be switched at the instruction clock rate of 20 ns. The MVP has been integrated into the MediaStation 5000, programmable multimedia system, described in detail in [L+94].

The key function of the system is MPEG compression. The data flow in the system during MPEG compression is shown in Figure 9.8. Video data is captured into the video buffer at a resolution of 320×240. The MVP reads the data from the video buffer and stores it in the main (DRAM) memory. The MVP

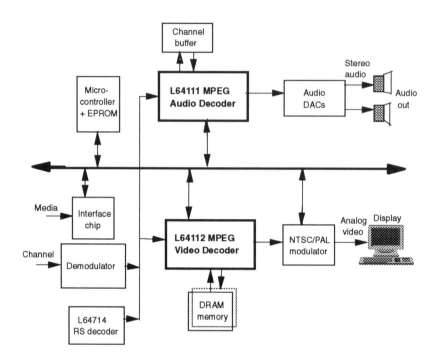

Figure 9.6 Block diagram of a typical system which uses LSI Logic's MPEG chip set. Courtesy of LSI Logic Inc.

performs all the compression functions on the data stored in the main memory. Similar operations are performed on the digitized audio samples. Once when the MVP completes the compression of a video or audio frame, the compressed bit stream is sent to the host computer, where the audio and video streams are multiplexed together, synchronized and stored on a disk or transferred to a network.

The performance results, reported in [L+94], show that the MediaStation 5000 system can achieve a real-time compression (30 frames/sec) of video sequences with resolutions of 320×240 pixels. The reported compression time for I frames is 17.7 ms, for P frames 27.3 ms, and for B frames 30.5 ms; for all frame types less than the video frame period of 33 ms.

9.2.4   IIT Vision Processor

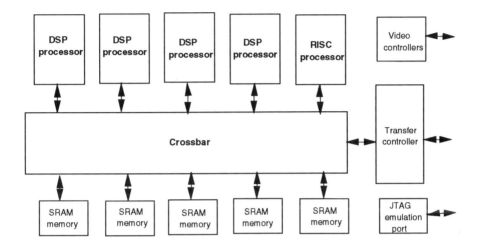

Figure 9.7 Block diagram of the TI's Multimedia Video Processor [GGV92].

Similarly to TI's MVP, Integrated Information Technology (IIT) has developed the Vision Processor (VP), a single-chip, highly-parallel, microcode-based, video signal processor [IIT92]. The VP can execute all of the JPEG, MPEG, and H.261 video compression and decompression algorithms. The block diagram of the Vision Processor is given in Figure 9.9.

The functions provided by VP include FDCT and IDCT, quantization and inverse quantization, zig-zag encoding and decoding, motion estimation and compensation, and filtering. The VP is controlled by an external host processor. For JPEG applications, the VP may be a high-performance general-purpose microprocessor, while for MPEG and H.261 applications, the IIT Vision Controller (VC) chip is used to handle pixel interface and frame buffer control functions.

IIT has also developed a Desktop Video Compression (DVC) board, which can be plugged into an IBM-compatible PC and used in various multimedia applications. The DVC board accepts a composite TV signal (NTSC or PAL), digitizes the video, compresses it, and saves the compressed bit stream on the hard disk. The board can also read the compressed data from the disk, decompress it, and display the video on a composite (NTSC or PAL) monitor. The block diagram of the DVC board is shown in Figure 9.10.

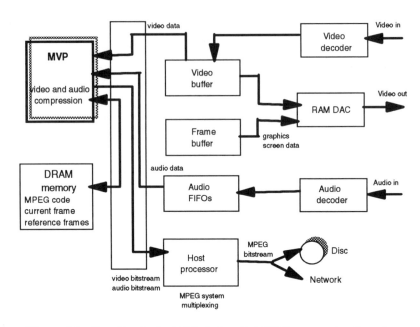

Figure 9.8   Data flow in the MVP during the MPEG compression [L+94].

The DVC board contains video capture and scaling circuitry, a video compression and decompression subsystem consisting of two video processors and one video controller, and video scaling and output circuitry.

9.3   VIDEO COMPRESSION SOFTWARE

In this section we describe Intel Indeo, which is a software solution for video compression [IV94].

Indeo video is a set of software compression techniques which together form a compressor/decompressor used for recording and compressing video data. For capturing and compressing video in real time, Indeo video runs on a video capture board (such as Intel Smart Video Recorder), and can compress a one-minute video clip from 50 MB to 9 MB and store on a hard disk.

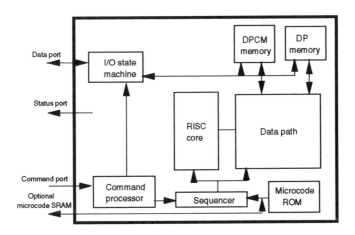

Figure 9.9 Block diagram of the IIT's Vision Processor (VP). Courtesy of Integrated Information Technology, Inc.

Indeo video allows playback video clips without requiring special hardware, which reduces the cost of decompression. The video file can be played back from any storage media. Table 9.6 illustrates video playback performance of Indeo video for three Intel microprocessors [IV94]. Note that Pentium processor is capable of playing back 320×240 frames at real-time speed, and 640×480 frames at half real-time speed.

Indeo video compression algorithm uses both lossy and lossless compression techniques, achieving compression ratio in the range from 6:1 to 10:1. Indeo video uses the following techniques to achieve the best image quality: color subsampling, pixel differencing, run length encoding, vector quantization, and variable content encoding.

Color subsampling is a lossy compression technique that reduces data by lowering the color depth of each pixel, compressed in bits/pixel. Using groups of 16 pixels, color subsampling does not change the Y component, but it uses only 8 bits to describe the color information for all 16 pixels of the U and V components. This process does reduction from 24 bits/pixels to an average of 9 bits/pixel, as illustrated in Figure 9.11.

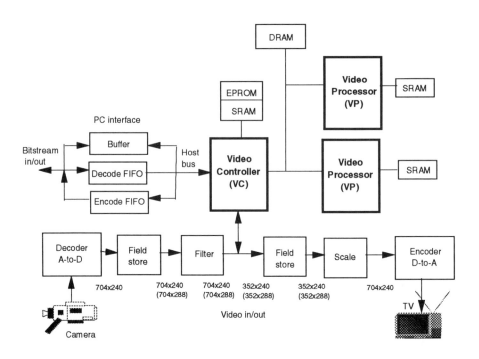

Figure 9.10 The block diagram of the IIT's Desktop Video Compression (DVC) board. Courtesy of Integrated Information Technology, Inc.

Pixel differencing is the operation that eliminates redundant data between two subsequent frames by comparing their pixels. It subtracts their values and saves only the information that is different.

Run length encoding (RLE) is a lossless compression technique which searches for consecutive pixels with the same color values. In Indeo system, the RLE technique is used after pixel differencing operation only when the resulting difference value is zero. In that case, RLE encodes the zero value and a count of consecutive zero values.

Vector quantization (VQ) is used when the difference value between adjacent pixels is not zero, but very close to zero. The VQ technique groups these differences together and then assigns an approximate value obtained from a table.

	Frame rate achieved [frames/sec (fps)]		
Micro-processor	**160x120 pixels**	**320x240 pixels**	**640x480 pixels**
i486 SX-25	30 fps	12 fps	2 fps
i486 DX2-66	30 fps	28 fps	10 fps
Pentium	30 fps	30 fps	15 fps

Table 9.6   Video playback performance of Indeo video [IV94].

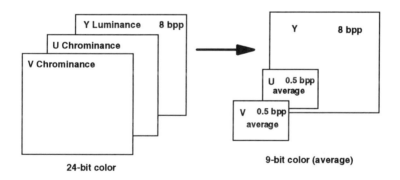

Figure 9.11   Color subsampling in Indeo video.

Variable content encoding (VCE) is a statistical compression technique (similar to Huffman coding), which further compresses the difference values by assigning variable length codes to them.

Indeo video is scalable, which means that when a video clip is played back, it determines what hardware is available and automatically adapts the playback

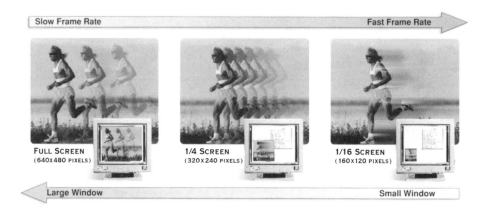

Figure 9.12 The tradeoff between the frame rate and the window size in Indeo. Courtesy of Intel Corporation.

quality for that configuration. The system is able to adjust the frame rate at which the file is played back in order to give the best perceived quality. The user can also make a tradeoff between the frame rate and window size, depending on which aspect is more important for the application, as illustrated in Figure 9.12.

9.4 GENERAL-PURPOSE PROCESSORS' SUPPORT FOR MPEG DECODING

Three advanced processors, specifically designed to support real-time MPEG decoding and some other multimedia functions, are: UltraSPARC V9 (Sun), PA-RISC 7100LC (Hewlett-Packard), and PowerPC620 (Motorola).

Architectural features of the PA-7100LC processor are briefly described in Section 7.4. The PowerPC620 processor is specifically designed for distributed multimedia systems. The processor supports both CPU-intensive and I/O-intensive applications. According to its designers, the processor can be used both for multimedia servers and multimedia workstations [Y+95].

The features of the UltraSPARC processor are described next.

UltraSPARC V9 and Its Visual Instruction Set

The latest version of Sun's SPARC processor implements a Visual Instruction Set (VIS), which provides some graphics and image processing capabilities needed for MPEG decoding [K+95]. The Visual Instruction Set supports new data types used for video images: pixels and fixed data. Pixels consist of four 8-bit unsigned integers contained in a 32-bit word, while fixed data consists of either four 16-bit fixed point components, or two 32-bit fixed point components both contained in a 64-bit word.

The VIS instructions are classified into the following five groups: conversation instructions, arithmetic and logical instructions, address manipulation instructions, memory access instructions, and a motion estimation instruction. Some of these groups are briefly described next.

Conversation instructions convert between various formats. For example, one visual instruction can convert 4 signed 16-bit pixels to 4 unsigned 8-bit pixels with clipping, and then pack them into one 32-bit word, which will usually take more than 12 operations with common instructions [G+95].

Arithmetic and logical instructions can process four pixels in one operation, performing either multiplication, addition, subtraction, or logical evaluations.

The motion estimation instruction, PDIST, computes the sum of the absolute differences between two 8 pixel vectors, which would require about 48 operations on most processors. Accumulating the error for a 16x16 block requires only 32 PDIST instructions; this operation typically requires 1500 conventional instructions.

The implementation of the PDIST (pixel distance) instruction is shown in Figure 9.13. The circuitry consists of three 4:2 adders, two 11-bit adders, and a 53-bit incrementer. It operates on 8-bit pixels, stored in a pair of double-precision registers, and produces the result in a single-cycle operation.

In general, the Visual Instruction Set provides four times or higher speed up for most of the time-critical computations for video decompression, including IDCT, motion compensation, and color conversion. According to an analysis, reported in [Z+95], the UltraSPARC processor is capable of decoding MPEG-2 video of 720x480 pixels resolution at 30 frames/sec entirely in software.

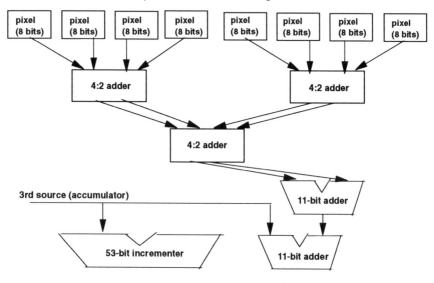

Figure 9.13 The implementation of the motion estimation instruction in SuperSPARC processor [K+95].

10

APPLICATIONS OF COMPRESSION SYSTEMS

In Chapter 3 we presented several multimedia applications, such as video conferencing and on-demand multimedia services, where compression plays an important role. In this chapter we describe several other applications of image and video compression systems.

10.1 JPEG APPLICATIONS

The most popular use of JPEG image compression technology include its use in photo ID systems, telecommunications of images, military imaging systems, and distributed image management systems. Many companies have developed JPEG-based systems for those applications. Several systems are summarized next [PM93].

The State of California Department of Motor Vehicles (Cal DMV) uses image technology in the *production of driver licenses and identification cards.* All drivers' pictures are stored as JPEG compressed images.

Autoview Inc. has developed a *vehicle image data storage system* using Telephoto's ALICE JPEG compression, which allows a storage of color images of vehicles for buyers or insurance-claim personnel.

Discovery Technologies, Inc. has developed several JPEG-based systems for medical applications. FilmFAX is a *high-resolution teleradiology system* using ALICE JPEG image compression. Tel-Med is a *microscope-based image com-*

munication system for transmitting high-resolution, full-color medical (pathology) images that also uses ALICE JPEG compression.

Moore Data Management Services has developed a JPEG-based *real-estate photo-display software package* that displays photos of real estate.

Tribune Solutions, a R&D division of Trubine Publishing Co., has developed PhotoView, an *image-archiving software package*. It uses ALICE JPEG compression.

XImage Corporation has developed image database systems, based on ALICE JPEG compression, specifically designed for *law enforcement* and related applications.

Distributed Image Management System

A distributed image management system, described in [A+90], can be used for various image applications. The block diagram of such a system is shown in Figure 10.1. The system consists of (a) image capture, compression, and transmission subsystems, (b) central and regional storage, and (c) local and remote displays.

In this system, images are captured, compressed, and then transmitted to regional databases via 19.2 Kbaud modems. A typical image (400 KB), compressed at 30:1 (to 13 KB), would be transmitted in 6-8 seconds. Compressed images are stored in regional databases, and can be accessed locally or remotely via terminals, or forwarded to a central database in compressed form.

Local displays are connected via LAN, and a simple high-speed JPEG decompression system may be satisfactory. For a large number of local displays, or if the LAN bandwidth becomes overloaded, individual JPEG decompression is needed at each local display.

Remote displays can be connected to the regional database via high or low speed modems. If high speed modems are used, the compression requirements are the same as captive terminals. If a low speed modem (2400 baud) or radio is used, the transmission time will be increased to 60 seconds, allowing a low cost software JPEG decompression.

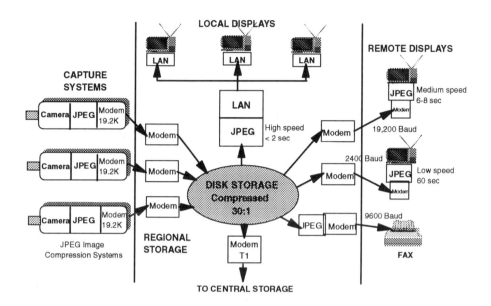

Figure 10.1 Block diagram of a distributed image management system which uses JPEG compression.

10.2 VIDEO COMPRESSION APPLICATIONS

Video compression standards (MPEG and px64) made feasible a number of applications summarized in Table 10.1. Table 10.1 specifies the required bandwidth for specific applications, applied compression standards, and typical frame size and frame rate.

In the world of desktop applications, four distinct applications of the compressed video can be summarized as: (a) consumer broadcast television, (b) consumer playback, (c) desktop video, and (d) videoconferencing [Ohr93a].

Consumer broadcast television, which includes home digital video delivery, typically requires a small number of high-quality compressors and a large number of low-cost decompressors. Expected compression ratio is about 5:1.

Consumer playback applications, such as CD-ROM libraries and interactive games, also require a small number of compressors with a large number of low-

MARKET	BANDWIDTH	STANDARD	SIZE	FRAME RATE [frames/sec]
Analog Videophone	5-10 Kbps	none	170x128	2-5
Basic video telephony	96-128 Kbps	px64 (H.261)	176x144 352x288	5-10
Basic confer encing	>= 384 Kbps	px64 (H.261)	352x288	15-30
Interactive multimedia	1-2 Mbps	MPEG-1	up to 352x288	15-30
Digital NTSC	3-10 Mbps	MPEG-2	720x480	30
HDTV	15-80 Mbps	MPEG-2 (FCC)	1200x800	30-60

Table 10.1   Applications of compressed video.

cost decompressors. However, the compression ratio is much greater (achieving 100:1) than that required for cable TV transmission.

Desktop video, which includes systems for authoring and editing video presentations, is a symmetrical application requiring the same number of decoders and encoders. The expected compression ratio is relatively small in the range 2-50:1.

Two-way videoconferencing also requires the same number of encoders and decoders, and expected compression ratio is about 100:1.

The VideoPhone 2500

The AT&T VideoPhone 2500 provides color, motion video over the analog loop on the public-switched telephone network [EKD93]. The VideoPhone 2500 uses a hardware codec for px64 compression algorithm, as illustrated in the block diagram in Figure 10.2.

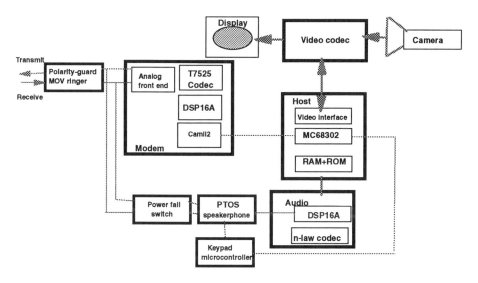

Figure 10.2 The architecture of the AT&T VideoPhone 2500 [EKD93].

The host processor is a MC68032 processor, while RISC-based peripheral processors handle serial communications. The modem provides full-duplex, synchronous data transmission at either 19.2 or 16.8 Kbits/s. Video codec, based on the Vision Processor from ITT (see Chapter 9), implements the H.261 (px64) compression standard. It is designed by Compression Labs. This block also contains host command data interface, and analog video input and output interfaces. Audio processor encodes and decodes audio data using a code-excited linear prediction (CELP) algorithm at 6800 bits/s.

The other components of the system are POTS and speakerphone for analog audio interface, telephone keypad, LCD display for color video images, and camera and lens.

The system multiplexes data from three independent sources: video, audio, and supervision information. The X.25 protocol is used for multiplexing. It creates virtual circuits for each channel, assigning different channel numbers to each circuit. The channel number serves as a prefix to each transmitted frame, and is used on the receiving side to demultiplex the data stream.

The allocated bandwidth for the VideoPhone is about 10 Kb/s, and its nominal frame rate is 10 frames/s. This gives an allocation of 1 Kb/s per frame. However, the number of bits per frame often exceeds this average, causing reduction in frame rate.

Digital Compression Camera

An interesting approach, that reduces the price of videoconferencing systems, integrates video compression hardware (in specific case the H.261 video compression standard) with the optics in a digital camera [BSS95]. Such a digital compression camera can be attached to the existing computers via a standard interface, such as SCSI.

The compression camera has been implemented using a set of SCSI commands, as illustrated in Figure 10.3. The compression camera device is controlled by a device driver implemented with the generic SCSI device driver. During the record/compression operation, a background daemon process, which is separated from the application process, reads the compressed video data using a SCSI command and stores the data in the shared memory. An application can access the compressed data from the shared memory using read calls.

The proposed compression camera has several advantages over the systems which use the separate camera and interface card. Besides the reduced cost, the compression camera provides the advantage regarding to the required bandwidth between the camera and the encoder. If the camera and the encoder are separated, the required bandwidth is typically over 150 Mb/s. Video compression reduces this data rate to about 1 Mb/s. Although various high speed proprietary interfaces for connecting low cost digital cameras to the desktop computers have been designed, no standard exists today. If the compression algorithm is integrated into the camera, no such standard is required, since the connection is internal.

On the other hand, as shown in Figure 10.3, a standard interface can be used to get the compressed video in the desktop computer. The SCSI (Small Computer System Interface) is a standard interface for attaching I/O devices (such as disc, tape, CD-ROM) to the desktop computers, and can accommodate the relatively low data rates received from the compression camera.

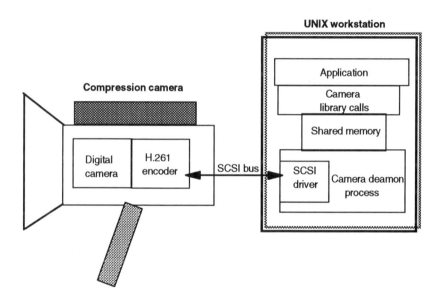

Figure 10.3 Compression camera interfaced to a standard workstation via SCSI interface [BSS95].

REFERENCES

[A+90]   D. R. Ahlgren, S. Acharya, J. Crosbie, and L. Roberts, "Applying
           Compression to Real World Image Management Systems", *JPEG
           Poster Session*, Boston, Mass., October 1990.

[A+93a]   B. D. Acidand, R. Aghevli, I. Eldumlati, A. C. Englander, and E.
           Scuteri, "A Video-Codec Chip Set for Multimedia Applications",
           AT&T Technical Journal, Vol. 72, No. 1, January/February 1993,
           pp. 50-65.

[A+93b]   R. Aravind, G. L. Cash, D. C. Duttweller, H.-M. Hang, B. G. Haskel,
           and A. Puri, "Image and Video Coding Standards", *AT&T Techni-
           cal Journal*, Vol. 72, January/February 1993, pp. 67-88.

[Ans93]   L. F. Anson, "Fractal Image Compression", *Byte*, October 1993, pp.
           195-202.

[BH93]   M. Barnsley and L. Hurd, "Fractal Image Compression", *AK Peters
           Ltd.*, Weelesley, MA, 1993.

[BSS95]   J. Back, F. Sung, and K. Severson, "An Object Based Architecture
           for a Digital Compression Camera", *Proc. of IEEE Compcon*, San
           Francisco, CA, March 1995, pp. 179-185.

[C+91]   L. G. Chen, W. T. Chen, Y. S. Jehng, amd T. D. Chiueh, "An
           Efficient Parallel Motion Estimation Algorithm for Digital Image
           Processing", *IEEE Transactions Circuits Systems*, Vol. 1, 1991, pp.
           378-385.

[C+94]   E. Chan, A. A. Rodriguez, R. Gandhi, and S. Panchanathan, "Ex-
           periments on Block-Matching Techniques for Video Coding", *Jour-
           nal of Multimedia Systems*, Vol. 2, No. 5, 1994, pp. 228-241.

[CHF94]   P. Chilamakuri, D. R. Hawthorne, and B. Furht, "Subband Image
           and Video Compression System and Its Comparison with JPEG
           and MPEG Systems", *Technical Report*, TR-CSE-94-44, Depart-
           ment of Computer Science and Engineering, Florida Atlantic Uni-
           versity, Boca Raton, FL, 1994.

213

[EKD93] S. H. Early, A. Kuzma, and E. Dorsey, "The VideoPhone 2500-Video Telephony on the Public Switched Telephone Network", *AT&T Technical Journal*, Vol. 72, January/February 1993, pp. 22-32.

[Fis94] Y. Fisher, "Fractal Compression: Theory and Applications to Digital Images". *Springer Verlag*, 1994.

[Fox91] E. A. Fox, "Advances in Interactive Digital Multimedia Systems", *IEEE Computer*, Vol. 24, No. 10, October 1991, pp. 9-21.

[Fri92] S. Friedel, "New Chip-Level VLSIs Implement Multimedia Standards", *Computer Technology Review*, Summer 1992, pp. 14-17.

[Fur94] B. Furht, "Multimedia Systems: An Overview", *IEEE Multimedia*, Vol. 1, No. 1, Spring 1994, pp. 47-59.

[Fur95] B. Furht, "A Survey of Multimedia Techniques and Standards - JPEG Compression", *Journal of Real-Time Imaging*, Vol. 1, No. 1, April 1995.

[Gha91] H. Gharavi, "Subband Coding Algorithms for Video Applications: Videophone to HDTV-Conferencing". *IEEE Transactions on Circuits and Systems for Video Technology*, Vol. 1, No. 2, June 1991, pp. 174-183.

[Gha92] M. Ghanbari, "An Adapted H.261 Two-Layer Video Codec for ATM Networks", *IEEE Trans. on Communications*, Vol. 40, No. 9, September 1992, pp. 1481-1490.

[Gil91] B. Gillman, "Getting the Best Out of Compression Products", *AV/Video*, March 1991.

[Gle93] W. E. Glen, "Digital Image Compression Based on Visual Perception and Scene Properties", *SPMTE Journal*, Vol.12, No. 5, May 1993, pp. 392-397.

[GGV92] K. Guttaj, R. J. Gove, and J. R. Van Aken, "A Single-Chip Microprocessor for Multimedia: The MVP", *IEEE Computer Graphics and Applications*, November 1992, pp. 53-64.

[GM90] H. Gharavi and M. Mills, "Block Matching Motion Estimation Algorithms - New Results", *IEEE Transactions Circuits and Systems*, Vol. 37, 1990, pp. 649-651.

[GT88] H. Gharavi and A. Tabatabai, "Subband Coding of Monochrome and Color Images", *IEEE Transacations on Circuits and Systems*, Vol. 35, February 1988, pp. 207-214.

[H+91]   K. Harney, M. Keith, G. Lavelle, L. D. Ryan, D. J. Stark, "The i750 Video Processor: a Total Multimedia Solution", *Communications of ACM*, Vol. 34, No. 4, April 1991, pp. 64-78.

[HHW93]   H.-C. Huang, J.-H. Huang, and J.-L. Wu, "Real-Time Software-Based Video Coder for Multimedia Communication Systems", *Journal of Multimedia Systems*, Vol. 1, 1993, pp. 110-119.

[HJS94]   M. L. Hilton, B. D. Jawerth, and A. Sengupta, "Compressing Still and Moving Images with Wavelets", *ACM/Springer-Verlag Journal of Multimedia Systems*, Vol. 2, No. 5, 1994, pp. 218-227.

[HM94]   A. C. Hung and T. H.-Y. Meng, "A Comparison of Fast Inverse Discrete Cosine Transfrom Algorithms", *Journal of Multimedia Systems*, Vol. 2, No. 5, 1994, pp. 204-217.

[IIT92]   IIT Vision Processor, Data Sheet, *Integrated Information Technology*, November 1992.

[IV94]   "Indeo Video: A Technical Overview", Intel Corporation, 1994.

[JJ81]   J. R. Jain and A. K. Jain, "Displacement Measurement and its Application in Interframe Image Coding", *IEEE Transacations on Communications*, Vol. 29, 1981, pp. 1799-1808.

[JV93]   "JView Evaluation Kit User's Guide", *LSI Logic*, Milpitas, CA, July 1993.

[K+81]   J. Koga, K. Iinuma, A. Hirano, Y. Iijima, and T. Ishiguro, "Motion Compensated Interframe Coding for Video Conferencing", *Proc. of the National Telecommunications Conference*, pp. G5.3.1-5.3.5, 1981.

[K+95]   L. Kohn, G. Maturana, M. Tremblay, A. Prabhu, and G. Zyner, "The Visual Instruction Set (VIS) in UltraSPARC", it Proc. of the IEEE Compcon, San Francisco, CA, March 1995, pp. 462-469.

[KAF94]   M. Kessler, J. Alexander, and B. Furht, "JPEG Still Image Compression Algorithm: Analysis and Implementation", *Technical Report TR-CSE-94-07*, Florida Atlantic University, Boca Raton, Florida, March 1994.

[L+94]   W. Lee, Y. Kim, R. J. Gove, and C. J. Read, "Media Station 5000: Integrating Video and Audio", *IEEE Multimedia*, Vol. 1, No. 2, Summer 1994, pp. 50-61.

[Lee95]   R.B. Lee, "Realtime MPEG Video via Software Decompression on a PA-RISC Processor", *Proc. of the IEEE Compcon*, San Francisco, CA, March 1995, pp. 186-192.

[LeG91]   D. LeGall, "MPEG: A Video Compression Standard for Multimedia Applications", *Communications of the ACM*, Vol. 34. No. 4, April 1991, pp. 45-68.

[LH91]   M. Liebhold and E. M. Hoffert, "Toward an Open Environment for Digital Video", *Communications of ACM*, Vol. 34, No. 4, April 1991, pp. 103-112.

[LK90]   A. S. Lewis and G. Knowles, "Video Compression Using 3D Wavelet Transforms", *Electronic Letters*, Vol. 26, No. 6, March 15, 1990, pp. 396-398.

[Lio94]   M. Liou, "Overview of the Px64 Kbit/s Video Coding Standard", *Comm. of the ACM*, Vol. 34, No. 4, April 1991, pp.59-63.

[MF94a]   P. Monnes and B. Furht, "Parallel JPEG Agorithms for Still Image Compression", *Proc. of Southeastcon '94*, Miami, Florida, April 1994, pp. 375-379.

[MF94b]   K. Morea and B. Furht, "Comparison of JPEG and Fractal Image Compression Techniques", *Technical Report*, Department of Computer Science and Engineering, Florida Atlantic University, Boca Raton, FL, 1994.

[Nel92]   M. Nelson, "The Data Compression Book", *M&T Books*, San Mateo, California, 1992.

[Ohr93a]   S. Ohr, "Video Teleconferencing Pushes Video Chips to Their Limit", *Computer Design*, February 1993, pp. 49-58.

[Ohr93b]   S. Ohr, "New Applications Driving Dedicated DSP Processors", *Computer Design*, Vol. 32, No. 5, May 1993, pp. 83-98.

[Pen93]   D. Y. Pen, "Digital Audio Compression", *Digital Technical Journal*, Vol. 5, No. 2, Spring 1993, pp. 28-40.

[PM93]   W. B. Pennenbaker and J. L. Mitchell, "JPEG Still Image Data Compression Standard", *Van Nostrand Reinhold*, New York, 1993.

[Poy94]   C. A. Poynton, "High Definition Television and Desktop Computing", chapter in the book "Multimedia Systems" by J.F. Koegel Buford, *ACM Press*, 1994.

[PSR93]   K. Patel, B. C. Smith, and L. A. Rowe, "Performance of a Software MPEG Video Decoder", *Proc. of the First International ACM Multimedia Conference*, Anaheim, CA, August 1993, pp. 75-82.

[Rau92]   I. R. Raucci, "JPEG Puts the Squeeze on PC Graphics", *Publish Magazine*, April 1992.

[Rip89]   G. D. Ripley, "DVI - A Digital Multimedia Technology", *Communications of the ACM*, Vol. 32, No. 7, July 1989, pp. 811-822.

[SR85]   R. Srinivasan and K. R. Rao, "Predictive Coding based on Efficient Motion Estimation", *IEEE Transactions on Communications*, Vol. 33, 1985, pp. 888-896.

[Ste94]   R. Steinmetz, "Data Compression in Multimedia Computing" - Standards and Systems", Part I and II, *Journal of Multimedia Systems*, Vol. 1, 1994, pp. 166-172 and 187-204.

[TSB95]   K. Tsunashima, J. B. Stampleton, and V. M. Bore, "A Scalable Motion-Compensated Subband Image Coder", *IEEE Transactions on Communications*, 1995.

[Vat84]   M. Vaterli, "Multi-dimensional Sub-Band Coding: Some Theory and Algorithms", *Signal Processing*, Vol. 6, April 1984, pp. 97-112.

[W+94]   K. S. Wang, J. O. Normile, H.-J. Wu, and A. A. Rodriguez, "Vector-Quantization-Based Video Codec for Software Only Playback on Personal Computers", *Journal of Multimedia Systems*, Vol. 2, No. 5, 1994, pp. 191-203.

[Wal91]   G. Wallace, "The JPEG Still Picture Compression Standard", *Comm. of the ACM*, Vol 34, No. 4, April 1991, pp. 30-44.

[Woo91]   J. W. Woods, "Subband Image Coding", *Kluwer Academic Publishers*, Boston, MA, 1991.

[Y+95]   J.K. Yuan, M.P. Taborn, D.C. Lee, and A. Tsay, " The PowerPC 620 Microprocessor in Distributed Computing", *Proc. of the IEEE Comcon*, San Francisco, CA, March 1995, pp. 308-314.

[Z+95]   C-G. Zhou, L. Kohn, D. Rice, I. Kabir, A. Jabbi, and X-P. Hu, " MPEG Video Decoding with the UltraSPARC Visual Instruction Set", *Proc. of the IEEE Comcon*, San Francisco, CA, March 1995, pp. 470-475.

QUESTIONS AND PROJECTS - PART II

QUESTIONS

1. Assume that an uncompressed multimedia presentation consists of:

    - 40 minutes of full-motion video ($640 \times 480 \times 24$ bits/frame @ 30 frames/sec),

    - 1,500 color images ($640 \times 320 \times 24$ bits/image), and

    - 1 hour of stereo sound (44 KHz/16 bits, 2 channels).

    a. Calculate the total memory capacity required to store this multimedia presentation.

    b. Propose compression algorithms for motion video and color images to compress this multimedia presentation.

2. Show an example of an image which is compressed using a JPEG progressive encoding algorithm which combines spectral selection and successive approximation techniques. The example should consist of 10 different scans. List all these 10 scans.

3. Draw a diagram which shows the organization of three types of pictures (frames) in MPEG compression algorithm: **I** (intrapictures), **P** (predicted pictures), and **B** (interpolated, bidirectional pictures).

    Describe how **I** and **P** are created. For **B** pictures, draw a diagram showing how these pictures are created. Show *the order of transmission* of these frames.

4. Draw the Differential Pulse Code Modulation (DPCM) compression encoder diagram and describe it. In JPEG compression, name *two* instances in which DPCM is applied.

5. A motion video consists of frames encoded according to CIF (Common Intermediate Format):

Luminance (Y) 288×352 pixels

Chrominance (Cb) 144 × 176 pixels

Chrominance (Cr) 144 × 176 pixels.

Assume that each pixel is 8 bits, and that the requirement for real-time video transmission is 30 frames/sec.

a. Calculate the total required *bit rate (in Mbits/sec)* to transmit this *uncompressed* motion video.

b. Assume that ISDN network is used for transmission of this motion video, whose capacity is px64 Kbits/sec, and $p = 10$. Calculate what *compression ratio* should be achieved by the px64 video compression technique in order to transmit the motion video in real-time.

c. If the QCIF (Quarter CIF) format is used instead the CIF format, and if the compression ratio of the applied compression algorithm is 40, calculate *the minimum required capacity* (in Kbits/sec) of the ISDN network which will be capable of transmitting this motion video in real-time. Also *find p* for this network.

6. The CCITT standard has defined a video compression scheme, H.261 (or px64), specifically for videoconferencing. Why other motion-video compression schemes, such as MPEG, are not suitable for this application?

7. Describe two JPEG progressive encoding techniques: spectral selection and successive approximation. What would be the reasons of applying JPEG progressive compressor rather than JPEG sequential compressor?

8. Explain the differences between asymmetric and symmetric applications regarding to compression and decompression. Give examples of both applications.

9. Describe and evaluate motion estimation techniques applied in MPEG and px64 compression algorithms.

10. Explain how the following features can be achieved using the MPEG compression algorithm: random access, fast forward/reverse searches, reverse playback, and editability.

11. Compare Motion JPEG (MJPEG) and MPEG compression techniques for full-motion video applications.

12. Evaluate JPEG and subband coding algorithms for image compression.

PROJECTS

13. **Software JPEG**

Project Assignment

Implement a software JPEG still image sequential compression algorithm, both encoder and decoder.

- Image to be compressed can be used from any available source (scaned, entered manually, or stored on a disc). The image can be either grayscale (8 bits), or color (24 bits), of any size.

- Quantization tables shall be created using the following C program:
  $$for(i = 0; i < N; i++)$$
  $$for(j = 0; j < N; j++)$$
  $$Q[i][j] = 1 + [(1 + i + j) \times quality];$$
  where *quality* is the quality factor from 1 to 25, with 1 specifying the highest quality and 25 the lowest one.

Presentation of Results

- Calculate the compression ratio, compressed bits/pixel, and root mean square (RMS) error for the following cases:
  Case a: use DC coefficients only,
  Case b: use DC, AC(0,1) and AC(1,0) coefficients,
  Case c: Quality = 1,
  Case d: Quality = 2,
  Case e: Quality = 10,
  Case f: Quality = 25.

- For all six cases, measure the execution times for both encoder and decoder as well as the total execution time.

- Write a well-documented report, which will include implementation details of the encoder and decoder, listing of the program, detailed results, and samples of the original and compressed images.

- Show the original image and decompressed images for different quality factors. Discuss the quality of the obtained images in terms of compression ratio and encoded bits/pixel.

14. **Motion vector estimation techniques**

Simulate and evaluate four motion vector estimation techniques (described in Section 7.1.1):

- Exhaustive search algorithm,
- Three-step-search algorithm,
- Two-dimensional logarithmic search algorithm, and
- Conjugate direction search algorithm.

Use three different cost functions:

- The Mean-Absolute Difference (MAD),
- The Mean-Square Difference (MSD),
- The Cross-Correlation Function (CCF).

Write a report, in which you will present and discuss obtained results.

PART III
Image and Video Indexing and Retrieval Techniques

11

CONTENT-BASED IMAGE RETRIEVAL

11.1 INTRODUCTION

As more and more image data are acquired and assume the role of "first-class citizens" in information technology, managing and manipulating them *as images* becomes an important issue to be resolved before we can take full advantage of their information content [CH92]. Image database and visual information system technologies have become major efforts to address this issue, and a variety of commercial systems are now available to accommodate different sizes of image resources, different computational platforms, and different commitments to investment. These systems have been used in an equally wide variety of applications, including medical image management, multimedia libraries, document archives, museums, transaction systems, computer-aided design and manufacturing (CAD/CAM) systems, geographic information systems, and criminal identification.

By and large these systems take a text-based approach to indexing and retrieval. That is, a text description, in the form of either key words or free text, is associated with each image in the database. These descriptions may be probed by standard Boolean database queries; and retrieval may be based on either exact or probabilistic match of the query text, possibly enhanced by thesaurus support [Caw93, AH+91]. Moreover, topical or semantic hierarchies may be used to classify or describe images using knowledge-based classification or parsing techniques; such hierarchies may then facilitate navigation and browsing within the database.[1] A good survey of such text-based query formation and matching techniques may be found in [ANAH93].

---

[1] This technology will be discussed in greater detail in Chapter 12.

However, there are several problems inherent in systems which are exclusively text-based. First, automatic generation of descriptive key words or extraction of semantic information to build classification hierarchies for broad varieties of images is beyond the capabilities of current machine vision techniques. Thus, these data must be keyed in by human operators. Apart from the fact that this is a time-consuming process when large quantities of images are involved, it is also very subjective. As a result, a retrieval may fail if a user forms a query based on key words not employed by the operator. It may also fail if the query refers to elements of image content that were not described at all by the operator and, hence, not represented by *any* key words. Moreover, certain visual properties, such as some textures and shapes, are difficult or nearly impossible to describe with text, at least for general-purpose usage.[2]

The alternative to relying on text is to work with descriptions based on properties which are inherent in the images themselves. The idea behind this is that the natural way to retrieve *visual* data is by a query based on the *visual* content of an image: the patterns, colors, textures, and shapes of image objects, and related layout and location information. For many applications such queries may be either supplemental or preferable to text, and in some cases they may be necessary. Furthermore, visual queries may simply be easier to formulate. For example, when it is necessary to verify that a trademark or logo has not been used by another company, the easiest way is to query an image database system for all images similar to the proposed pattern [ANAH93]. Similarly, given a library of fabric patterns, a fashion designer may want to retrieve specific images based on which colors are present and the texture in which they are mixed. Formulating a query for such a search would involve either selecting or creating one or more representative examples and then searching for images which resemble those examples. In both of these cases, search is driven by first establishing one or more *sample images* and then identifying specific *features* of those sample images which need to match images in the resource being searched. Because the focus is on those visual features which comprise the content of the images themselves, this technique is often called *content-based* image retrieval.

Queries based on image content require a paradigm which differs significantly from that of both traditional databases and text-based image understanding systems. First, such queries cannot be as logically rigorous as those expressed

---

[2] Highly specialized communities, such as, for example, surgical pathologists, may have their own idiosyncratic vocabulary for dealing with images related to their work (in this case tissue samples examined under a microscope); but general-purpose users are likely to be unfamiliar with such vocabularies and may even be confused by their use of familiar words in unfamiliar ways.

in text. Images are not atomic symbols, so symbol equality is not a particularly realistic predicate. Instead, searches tend to be based on *similarity*; and resemblance is more important than perfectly matching bit patterns. On the other hand all that is *similar* is not necessarily *correct*, so this paradigm tends to entail the retrieval of false positives which must then be manually discarded by the user. In other words the paradigm rejects the idea that queries may be expressed in terms of necessary and sufficient conditions which will determine *exactly* which images one wishes to retrieve. Instead, a query is more like an *information filter* [BC92], defined in terms of specific image properties, which reduces the user's search task by providing only a small number of candidates to be examined; and it is more important that candidates which are likely to be of interest are not *excluded* than it is that possibly irrelevant candidates be *included*. Consequently, as opposed to the usual artificial intelligence approach to scene analysis, there are no well-defined procedures which automatically identify objects and assign them to a small number of pre-defined classes [N+93].

(As an aside, it is worth observing that this rejection of logical necessity and sufficiency as criteria which guide a search is very much in sympathy with Ludwig Wittgenstein's similar rejection of these criteria as a basis for knowledge representation [Wit74]. Wittgenstein argued that knowledge of concepts was limited to the extent to which they shared "family resemblances." More recently, Gerald Edelman has demonstrated how we form perceptual categories on the basis of both local and global features of visual stimuli [Ede87]. He has subsequently argued that such perceptual categorization lays the groundwork for an understanding of "family resemblances" among both sensory stimuli and verbal concepts [Ede89]. Thus, the paradigm of queries based on image content is not only more feasible from an engineering point of view but may be further justified as being consistent with observed physiological properties.)

An important consequence of this paradigm is that functionality which supports interactivity between the user and the database through a *visual interface* is as significant as the ability to support image-based queries. Such interactivity enhances the user's ability to *express* queries, *evaluate* query results, and *refine* queries on the basis of those evaluations. Thus, the issues which must be addressed by a database which supports queries based on image content may be summarized as follows [N+93]:

- selection, derivation, and computation of image features and objects that provide useful query expressiveness;

- retrieval methods based on similarity, as opposed to exact matching;

- a user interface that supports the visual expression of queries and allows query refinement and navigation of results

- an approach to indexing which is compatible with the expressiveness of the queries.

Figure 11.1 illustrates a system architecture which supports this particular approach to content-based image retrieval. At the heart of this architecture is a database structured with respect to image features which are extracted for both data entry and query interpretation and compared for similarity during retrieval. The user interface then supports the "closing of the loop," which relates the formulation of queries to the browsing of retrieved data.

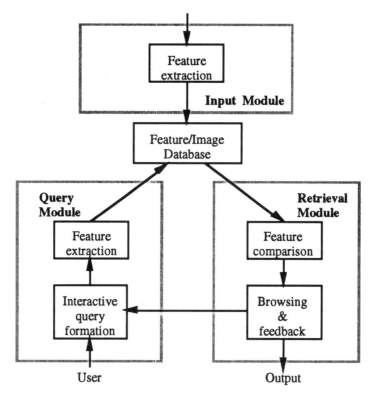

Figure 11.1   Architecture for content-based image retrieval.

There are two ways in which this paradigm based on image features may be approached. We call the first the *model-based* approach; and it is similar in nature to the artificial intelligence approach to scene analysis in that it assumes some *a priori* knowledge—the model—of how images are structured [CF82]. The second is the more *general-purpose* approach, which makes no assumptions at all about the nature of the images being searched. In the remainder of this Chapter, we shall concentrate on those features which have been demonstrated to be most effective for the general-purpose approach; but first it is worth making some observations about the model-based approach.

Clearly, the model-based approach is most effective when the model supports a rich collection of assumptions regarding the nature of the images being searched. An area which has been particularly conducive to the formulation of such assumptions has been face recognition. This is frequently a case where one does not have to worry about invariance of properties with respect to size, translation, and rotation. If one assumes a collection of face photographs, all shot from a common camera with uniform magnification, then one can make many powerful assumptions as to how an image may be segmented into components, each of which has a very specific structural description. Such a model readily translates into a " feature vector" which represents each image. Once those feature vectors have been defined, a variety of techniques based on their mathematical properties may be engaged to facilitate their efficient search [W+94].

The general-purpose approach requires a more general model of which features should be examined and how different instances of those features should be compared for proximity. On the basis of results which have been established to date, those features which have been seen to be most effective have been color, texture, shape of component objects, and relationships among edges which may be expressed in terms of line sketches. Each of these features will now be discussed in terms of how it may be most appropriately modeled to serve the needs of retrieval. Once we have established how these features may be modeled, we shall then discuss how they may be used to *index* images in a way which will improve the efficiency of the search process.

11.2   IMAGE FEATURES FOR CONTENT-BASED RETRIEVAL

A database which supports content-based image retrieval must also support two main data types: *objects* and *scenes*. A scene is an image which contains zero or more objects, and an object is a part of a scene. For instance, a car is an object in a traffic scene. Both of these data types need to be represented by their visual features: color, texture, shape, location and relationships among edges which may be abstracted as line sketches [CF82]. These visual features are more feasible to compute than semantic interpretation and have broad and intuitive applicability. We shall now discuss the derivation, calculation and comparison of these features, giving examples from their application in image retrieval.

11.2.1   Color

The Global Property of Color Distribution

The distribution of color is a useful feature for image representation, especially in textured images and other images which are not particularly amenable to segmentation. Color distribution, which is best represented as a histogram of intensity values, is more appropriate as a global property which does not require knowledge of how an image is composed of component objects. If we assume that the color of any pixel may be represented in terms of component red, green, and blue values, then a histogram may be defined, each of whose bins corresponds to a range of those values for each of the components. Such histograms are invariant under translation and rotation about the view axis and change only slowly under change of angle of view, change in scale, and occlusion [SB91]. Therefore, the color histogram is a suitable quantitative representation of image content.

Histograms may be compared by an *intersection* operation. Let Q and I be two histograms, corresponding, for example, to a query image and an image in a database. Assume that both histograms contain N bins, ordered in the same manner. The intersection of these two histograms is defined as follows:

$$\sum_{i=1}^{N} \min(I_i, Q_i) \qquad (11.1)$$

This sum may be interpreted as enumerating the number of pixels which are common to both histograms, which corresponds to the set-theoretic sense of intersection. This value may then be normalized by the total number of pixels in one of the two histograms. For example, if the query (Q) histogram is used for normalization, the normalized intersection would be defined as follows:

$$S(I,Q) = \frac{\sum_{i=1}^{N} \min(I_i, Q_i)}{\sum_{i=1}^{N} Q_i} \tag{11.2}$$

This value will always range between 0 and 1. Previous work has shown that this metric is fairly insensitive to change in image resolution, histogram size, occlusion, depth, and view point [SB91].

While this technique is quite simple, it can also be computationally expensive. If it is applied to searching a database by comparing the query image with all the database images, then the computational time cost is $O(NM)$, where N is the number of histogram bins and M is the total number of images in the database. If the database must be exhaustively searched, then the only way to reduce search time is to reduce the number of histogram bins. One method involves transforming a representation based on red, green, and blue axes into one based on *opponent color* axes, rg, by, and wb, whose intensities are defined in terms of red (R), green (G), and blue (B) intensities as follows [SB91]:

$$rg = R - G \tag{11.3a}$$
$$by = 2B - R - G \tag{11.3b}$$
$$wb = R + G + B \tag{11.3c}$$

The key advantage of this alternative is that the wb axis (also called "intensity") can be more coarsely sampled than the other two. This reduces the sensitivity of color matching to variation in the overall brightness of the image, while it also reduces the number of bins in the color histogram. If rg and by are divided into 16 sections, while wb is partitioned into only 8 sections, the resulting histogram will consist of 2048 bins. Experimental results have shown that such a coarse segmentation of color space is sufficient. Indeed, the evidence is that further refinement does not necessarily improve matching performance in image retrieval [SB91].

Another systematic way to quantize color space is to apply a clustering technique to the input data in order to determine the K "best" colors in a given color space. Standard algorithms, such as minimum sum of squares clustering [DH73], can be used for this purpose. The result of the clustering may then be interpreted as a partition of the color space into K "super-cells," each of which will correspond to a histogram bin. Color histograms of images or objects can then be calculated as the normalized count of the number of pixels that fall in each of these super-cells [N+93]. The advantage of this approach is that the clustering process will take into account the color distribution of images over the entire database; and this will minimize the likelihood of histogram bins in which no or very few pixels fall, thus resulting a very efficient quantization of the entire color space for that particular database population.

An alternative to clustering the color distribution is simply to recognize that only a small number of histogram bins tend to capture the majority of pixels of an image. Therefore, only the largest bins (in terms of pixel counts) need be selected as the representation of any histogram; and, as long as the bins of the query and image histograms are appropriately matched, intersection may be computed over this reduced set. Experiments have shown that such reduction does not degrade the performance of histogram matching. In fact, it may even enhance it, since small histogram bins are likely to be noisy, thus distorting the intersection computations. One experimental database has been achieving successful color-based retrieval results by reducing a histogram to its twenty most populated bins [G+94].

Figure 11.2 shows an example of the use of color histograms for image retrieval, based on querying a database of 400 images with an example image. The histogram consists of only the twenty most populated color bins. The retrieved images match the example image very well in terms of color content, showing the effectiveness of color distribution as a representation.

One major problem with histogram intersection is that all bins are treated entirely independently. There is no attempt to account for the fact that certain pairs of bins correspond to colors which are more perceptually similar than other pairs. Even clustering the color space into super-cells may give a better account of proximity *within* a cell; but similarity *between* cells is still not represented. If the Q and I histograms are regarded as N-dimensional vectors, then bin proximity may be represented as a (probably symmetric) $N \times N$ matrix A. The distance between Q and I may then be defined as follows [Iok89]:

$$D(I, Q) = (I - Q)^t A (I - Q) \qquad (11.4)$$

The proximity coefficients in A may be based on proximity is the $L^*u^*v^*$ space defined by the CIE (Commission Internationale de L'Eclairage) [Iok89]; or, perhaps in the interest of the restricted domain of a particular database of images, it may be determined directly from experimental data gathered from human visual perception studies. This formula has been demonstrated to achieve desirable retrieval performance [Iok89].

The biggest disadvantage of a histogram is that it lacks any information about location. This problem may be solved by dividing an image into sub-areas and calculating a histogram for each of those sub-areas. Increasing the number of

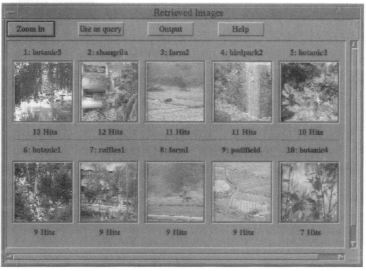

Figure 11.2 An example of image retrieval based on color histograms: The top twenty color bins are used in indexing and matching. The number of "hits" is the number of matching bins.

sub-areas increases the information about location; but it also increases the memory required to store histograms and the time required to compare them. A viable approach seems to be to partition an image into a 3×3 array of 9 sub-areas; thus, 10 histograms need to be calculated, one for the entire image and one for each of the 9 sub-areas [G+94]. A more localized technique involves the color analysis of individual objects, which will now be discussed.

The Local Property of Object Color

Apart from using color distribution to represent *scene* content, color, together with texture and geometric features, is also an important property in *object* representation. Since most natural objects do not have a single color, a simple and commonly used representation is an *average* value with respect to some color space. Retrieval of objects can then be based on these average colors. Here we shall focus our attention on color as a means of both identifying objects and describing them.

Object Detection by Color-Based Segmentation Color-based segmentation involves grouping adjacent pixels with similar color properties into common regions. The success of this technique depends on the properties which are considered. The most effective approach has been to compute proximity in HVC space [TC92]. This is yet another three-dimensional color space, whose components are known as Hue (the type of color), Value (also known as intensity or luminance), and Chroma (also called saturation—the degree to which color is present, as opposed to an achromatic gray).

Segmentation begins by dividing an image into achromatic and chromatic regions, based strictly on the Chroma component of each pixel. The Hue component is then used to further segment the image into a set of uniform regions based on a histogram difference metric [GZ94].[3] Finally, post-processing is carried out to recover from over-segmentation. Experimental results have shown that the performance of this approach is fairly good when it is applied to images with reasonably homogeneous regions, but the performance may be degraded in the presence of strong highlights or shading.

Region Detection by Pre-Defined Colors Despite extensive research into the development of robust and accurate segmentation algorithms, the results have been rather limited, particularly when attempts are made to apply them to broad and general image domains. Influences of shading and complicated illumination conditions may lead to many erroneous segments. Part of the problem is that most segmentation processes group pixels of similar properties into a common region without any knowledge of the type of object associated with that region. Such knowledge is only possible if the domain is limited by *a priori* assumptions; but, in that case, those assumptions will also carry information about limitations in region color.

---

[3] Histogram difference metrics will be reviewed in Chapter 12.

Object name	Hue	Chroma
Skin	0.7–1.2	> 20.0
Red apple	0.0–0.6	> 20.0
Lawn/Forest	1.5–3.0	> 20.0
Sunny sky	3.8–4.4	> 20.0
Yellow lemon	0.9–1.4	> 20.0
Gray (concrete)	–	< 10.0

Table 11.1 Color ranges of some pre-defined objects.

While it is not practical to try to develop an all-purpose set of assumptions about object colors, it is still possible to define a limited set of assumptions for common objects such as human faces, sunny skies, the sea, forests, and lawns. Those objects which may best be so defined are objects for which the hue falls into a narrow characteristic range, despite variations in illumination and photographing conditions in different scenes. Some of these ranges are listed in Table 11.1. Because such a table defines intervals in HC space, independent of V values, it provides an alternative approach to segmentation which performs classification at the same time, simply by identifying regions of contiguous pixels, all of which match one of its entries. Figure 11.3 shows an example of object detection applying this approach, in which the lawn area is correctly detected based on the color range defined in Table 11.1.

11.2.2 Texture

Texture has long been recognized as being as important a property of images as is color, if not more so, since textural information can be conveyed as readily with gray-level images as it can in color. Nevertheless, there is an extremely wide variety of opinion concerned with just what texture is and how it may be quantitatively represented [TJ93]. The texture analysis techniques which have been developed and are most extensively used involve two classes of methods: structural and statistical. The former describe texture by identifying structural primitives and their placement rules, but these methods tend to be most effective when applied to textures which are too regular to be of any practical interest. Statistical methods characterize texture in terms of the spatial distribution of image intensity, just as color analysis is concerned with the distributions of the intensities of components of a color space. Statistical texture

Figure 11.3 An image to be segmented on the basis of pre-defined object colors.

Figure 11.3 *(continued)*
Lawn is distinguished from other objects in the scene.

features may be derived from Fourier power spectra, co-occurrence matrices, Markov random field (MRF) models, fractal models, and Gabor models [TJ93].

Given such a wide scope of variety, it is not practical to investigate exhaustively the applicability of each of these models. Instead, this discussion will focus on three of the more popular approaches which have been demonstrated to be particularly effective in image retrieval: Tamura features [TMY76], the Simultaneous Autoregressive (SAR) model [KK87], and Wold features [LP94]. These methods are all fairly simple to implement and have been shown to be useful for texture classification, segmentation, and synthesis, the three primary applications of texture analysis discussed in [TJ93]. More specifically, they have been tested against the entire database of Brodatz texture images [Bro66], a validation technique which is now generally accepted as a testing standard [PKL93]. These models will now be summarized, followed by a discussion of how they may be compared for proximity.

Techniques

Tamura Features The Tamura features were intended to correspond to properties of a texture which are readily perceived [TMY76]. They are called *contrast*, *directionality*, and *coarseness*. The computation of each of these features will now be reviewed.

Contrast The quantification of contrast is based on the statistical distribution of pixel intensities. More specifically, it is defined in terms of the kurtosis $\alpha_4 = \frac{\mu_4}{\sigma^4}$ of the distribution (where σ is the standard deviation and μ_4 is the fourth moment about the mean). The formula for computing contrast is as follows:

$$F_{con} = \frac{\sigma}{\alpha_4^{1/4}} \qquad (11.5)$$

This provides a global measure of the contrast over the entire image.

Coarseness Coarseness is a measure of the granularity of the texture. It is based on moving averages computed over windows of different sizes, $2^k \times 2^k$, where k usually varies between 0 and 5. One of these k values gives a "best fit" and is used to compute the coarseness value.

At pixel (x, y) the moving average over a window of size $2^k \times 2^k$ is given by the following formula:

$$A_k(x, y) = \sum_{i=x-2^{k-1}}^{x+2^{k-1}-1} \sum_{j=y-2^{k-1}}^{y+2^{k-1}-1} \frac{g(i, j)}{2^{2k}} \qquad (11.6)$$

$g(i, j)$ is the gray-level at pixel (i, j). Differences between the moving averages in the horizontal and verical directions are computed as follows:

$$E_{k,h}(x, y) = |A_k(x + 2^{k-1}, y) - A_k(x - 2^{k-1}, y)| \quad \text{(11.7a)}$$
$$E_{k,v}(x, y) = |A_k(x, y + 2^{k-1}) - A_k(x, y - 2^{k-1})| \quad \text{(11.7b)}$$

At each pixel, the value of k which yields the overall maximum for $E_{k,h}(x, y)$ and $E_{k,v}(x, y)$ is used to set the following *optimization* parameter:

$$S_{best}(x, y) = 2^k \quad \text{(11.8)}$$

Given an $m \times n$-pixel image, the coarseness metric is then computed from the optimization parameter as follows:

$$F_{crs} = \frac{1}{m \times n} \sum_{i=1}^{m} \sum_{j=1}^{n} S_{best}(i, j) \quad \text{(11.9)}$$

The computation of this measure can be modified to apply to small objects or regions. Also, in order to reduce the amount of computation for this measure, a lower image resolution can be used with fewer window sizes.

Directionality To compute the directionality measure, a gradient vector is calculated at each pixel. The magnitude and angle of this vector are defined, respectively, as follows:

$$|\Delta G| = (|\Delta_H| + |\Delta_V|)/2 \quad \text{(11.10a)}$$
$$\theta = \arctan(\Delta_V/\Delta_H) + \frac{\pi}{2} \quad \text{(11.10b)}$$

The horizontal and vertical differences Δ_H and Δ_V are computed over a 3×3 array centered on the pixel. Once the gradients have been computed for all pixels, a histogram of θ values, H_D, is constructed by first quantizing θ into some given set of bins and then, for each bin, counting those pixels for which the corresponding magnitude $|\Delta G|$ is larger than a given threshold. This histogram will exhibit strong peaks for highly directional images and will be comparatively flat when the images are non-directional. This technique may be modified to deal with heterogeneous images. The entire histogram may then be summarized by an overall directionality measure based on the sharpness of the peaks as follows:

$$F_{dir} = \sum_{p}^{n_p} \sum_{\phi \in w_p} (\phi - \phi_p)^2 H_D(\phi) \quad \text{(11.11)}$$

In this sum p ranges over the n_p peaks; and for each peak p w_p is the set of bins over which it is distributed, while ϕ_p is the bin in which it assumes its highest value. The only peaks included in this sum must have predefined ratios between the peak and its surrounding valleys.

SAR Model   Given an image of gray-level pixels, the SAR model provides a description of each pixel in terms of its neighboring pixels [KK87]. The class of SAR models is a subset of the class of conditional Markov (CM) models. Such a model has the advantage of requiring fewer parameters than a CM model; and it can also be extended to include simultaneous moving average (SMA) and simultaneous autoregressive moving average (SARMA) models [TJ93]. Suppose an image is an $M \times M$ array of pixels. Let $s = (s_1, s_2)$ be one of these pixels, where s_1 and s_2 both range between 1 and M. Then the SAR value for s is defined by the following formula:

$$g(s) = \mu + \sum_{r \in D} \theta(r)g(s + r) + \epsilon(s) \tag{11.12}$$

The parameters for this formula may be summarized as follows:

- μ is a *bias* value which depends on the mean gray-level of the entire image.

- D defines a *neighbor set*; it is commonly chosen as the second-order neighborhood illustrated in Figure 11.4.

- $\theta(r)$ is a set of *model parameters* which characterize the dependence of pixel s on the members of its neighbor set r; *symmetric* model parameters satisfy the property $\theta(r) = \theta(-r)$.

- $\epsilon(s)$ is an independent Gaussian random variable with zero mean whose variance σ^2 models the noise level.

$$
\begin{array}{ccc}
(-1,1) & (0,1) & (1,1) \\
(-1,0) & (0,0) & (1,0) \\
(-1,-1) & (0,-1) & (1,-1)
\end{array}
$$

Figure 11.4   The second-order neighborhood of the pixel at (0,0).

Perceptually, σ measures the *granularity* (or "business") of the texture: a higher variance implies a finer granularity or less coarseness . The θ parameters provide a means to represent directionality ; for example, a vertically oriented texture will be modeled by higher values for $\theta(0, 1)$ and $\theta(0, -1)$. Needless to say, the effectiveness of the SAR model depends on the quality of estimates for θ and σ. These are usually determined by either least squares estimation (LSE) or maximum likelihood estimation (MLE) [KK87].

RISAR Model   The basic SAR model is not rotation invariant: Two images of the same texture which differ only by rotation will have different model parameters. A rotation-invariant variation of the model (RISAR) has been defined to obtain rotation-invariant texture features [MJ92]. The model value at a pixel is now described in terms of a weighted sum of its neighbors lying on circles of different radii centered at the pixel:

$$g(s) \;=\; \mu + \sum_{i=1}^{p} \theta_i x_i(s) + \epsilon(s) \tag{11.13a}$$

$$x_i(s) \;=\; \frac{1}{8i} \sum_{r \in N_i} w_i(r) g(s + r) \tag{11.13b}$$

The additional parameters for this model are as follows:

- p is the number of circles which are used to describe the neighborhood.

- N_i is the set of neighbor pixels associated with the ith circle around s. (This is illustrated in Figure 11.5.)

- w_i is a set of fixed, pre-computed weights indicating the respective contributions of the neighbor pixels in the ith circle.

Once again, θ and σ are estimated using LSE or MLE. The critical element of this model is the placement of the circles. As p gets larger, computation becomes more time-consuming; but if p is too small, the effect of rotation-invariance will not be achieved. Experimental results have established a value of 2 for p (with neighborhoods defined as in Figure 11.5) to be satisfactory [KZL94].

MRSAR Model   Another variation on the SAR model is a *multiresolution* (MRSAR) model [MJ92] which represents textures at a variety of resolutions in order to describe textures of different granularities. In this case the image is represented by a multiresolution Gaussian pyramid obtained by low-pass filtering and subsampling operators applied at several successive levels. Either the SAR or RISAR model may then be applied to each level of this pyramid, as is illustrated in Figure 11.6. As is the case with RISAR, determining a

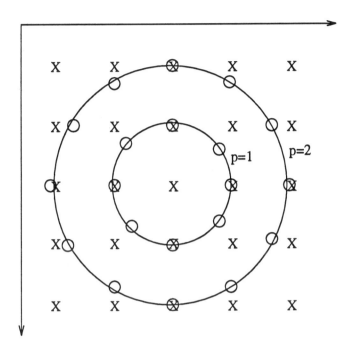

2-variate RISAR

X : pixel locations

O : points interpolated from neighbouring pixels

Figure 11.5 Neighborhoods for RISAR computation.

number of layers which is both effective and computationally realistic is a key problem. Experimental results have shown that four pyramid layers can yield high classification accuracy [KZL94].

Wold Features   Another approach to the representation of texture is based on the decomposition of a two-dimensional regular and homogeneous random field. If $\{y(n,m)\}$ is such a field, then it has the following unique orthogonal decomposition [SFP94]:

$$y(n,m) = w(n,m) + v(n,m) \tag{11.14}$$

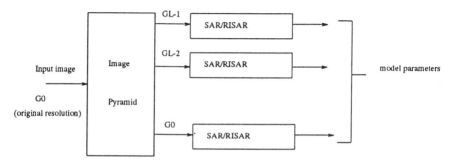

Figure 11.6   Multiresolution SAR/RISAR model.

$\{w(n, m)\}$ is a purely indeterministic field with a moving average representation [LP94]:

$$w(n, m) = \sum_{(0,0) \preceq (k,l)} a(k, l) u(n - k, m - l) \qquad (11.15)$$

where the innovation field $u(n, m)$ is white, $a(0, 0) = 1$ and $\sum_{(0,0) \preceq (k,l)} a^2(k, l)$ is finite.

$\{v(n, m)\}$ is a deterministic random field which may be further decomposed with respect to a family of non-symmetric half-plane total-order definitions where the angle of the dividing slope θ is rational (i.e. $\tan \theta = \beta/\alpha$, where α and β are coprime integers). Let O be the set of all these divisions. Then the decomposition of the $\{v(n, m)\}$ field is given by:

$$v(n, m) = p(n, m) + \sum_{(\alpha,\beta) \in O} e_{(\alpha,\beta)}(n, m) \qquad (11.16)$$

$\{e_{(\alpha,\beta)}(n, m)\}$ is the *evanescent* field corresponding to the division of the non-symmetric half-plane; and $\{p(n, m)\}$ is a *half-plane deterministic* random field. The purely indeterministic, half-plane deterministic (or *harmonic*), and evanescent fields are sometimes called the *Wold components* of the random field $\{y(n, m)\}$ [LP94].

This decomposition provides another approach to describing textures in terms of perceptual properties. In this case the properties are *periodicity*, *directionality*, and *randomness*. Each of these properties is associated with dominance of a different Wold component: Periodic textures have a strong harmonic component, highly directional textures have a strong evanescent component, and less structured textures tend to be strongest in the indeterministic component. These correlations have been established with both the Brodatz texture database and collections of images taken under more natural circumstances [LP94].

Proximity Comparison

Each of these models provides a representation of image texture as a vector of numbers. Just as we addressed the comparison of histograms in Section 11.2.1, we must also provide a quantitative metric of similarity between two such texture vectors. There are several ways in which such a metric may be defined. Let us assume that texture is being represented by two J-dimensional vectors X^1 and X^2. The simplest way to compare their similarity would be by a *Euclidean* distance metric :

$$D_{euclid} = \sum_{j=1}^{J} (X_j^1 - X_j^2)^2 \qquad (11.17)$$

This assumes that all dimensions are independent and of equal importance. The *Mahalanobis* distance metric allows the representation of assumptions of interdependence among the dimensions and relative weights of importance. All this information is captured in a $J \times J$ *covariance matrix*. If C is the covariance matrix for the range of feature values from which the vectors X^1 and X^2 have been constructed, then their Mahalanobis distance is defined as follows:

$$D_{mahal} = (X^1 - X^2)^T C^{-1} (X^1 - X^2) \qquad (11.18)$$

There is also a *simplified* Mahalanobis distance which assumes independence of the feature values. In this case it is only necessary to compute a variance value for each feature, c_j. The simplified Mahalanobis distance is then defined as follows:

$$D_{simpl} = \sum_{j=1}^{J} \frac{(X_j^1 - X_j^2)^2}{c_j} \qquad (11.19)$$

Further discussion of these distance metrics may be found in [DH73].

Experiments with the Brodatz Database

As has already been observed, the Brodatz database is the *de facto* standard for evaluating texture algorithms. The database contains 112, 512×512 8-bit gray-level images, each of which represents a different texture class. Evaluation is based on the following procedure [PKL93]: Nine 128×128 sub-images are extracted from the center of each Brodatz image. Features are computed by applying the model to be evaluated to each of these 1008 sub-images. The distance of each test sub-image to all the other sub-images is then computed; and the *retrieval rate* is defined to be the percentage of the n nearest sub-images which belong to the same class as the test sub-image. The *average retrieval rate* is then computed by using each of the 1008 sub-images as a test sub-image, and this value may be plotted against different values of n. Figure 11.7 [KZL94] shows such a plot for a series of experiments in which the values of n were 8, 13, 18, 23, 28, 33 and 38. Each curve corresponds to the evaluation of a different model. A higher curve indicates better performance. The motivation for using this testing procedure is that, as was discussed in Section 11.1, the user will generally be interested in finding the n most similar patterns to the query pattern and having those patterns displayed.

SAR on Brodatz When applying the SAR model on the Brodatz database, several issues have to be considered. These include [KZL94]:

Choice of Neighborhood: A symmetric second-order neighborhood, as in Figure 11.4 was assumed. A larger neighborhood causes deterioration in performance due to the averaging effect of more parameters. This choice yielded five model parameters per resolution level.

Estimation Scheme: The maximum likelihood (MLE) method was found to be more accurate though it required slightly more computation than the least squares (LSE) method (normally only 2 iterations per parameter).

Choice of Distance Function: The Mahalanobis distance function gave the best results since it takes into account the statistics of the data (in the covariance matrix). The simplified Mahalanobis distance, though easier to compute, was not so accurate, since the covariance matrix was in general not diagonal.

Number of levels for MRSAR: Four levels of resolution were chosen for the MRSAR model, i.e. 128×128, 64×64, 32×32 and 16×16 resulting in twenty model parameters overall. The multiresolution model performs better than the simple SAR (see Figure 11.7) since it incorporates the dependencies between pixels at various scales (rather than choosing a large neighborhood at a single resolution).

Rotation invariant (RISAR) features: As expected, the RISAR model performed better when the database was expanded to include rotated ver-

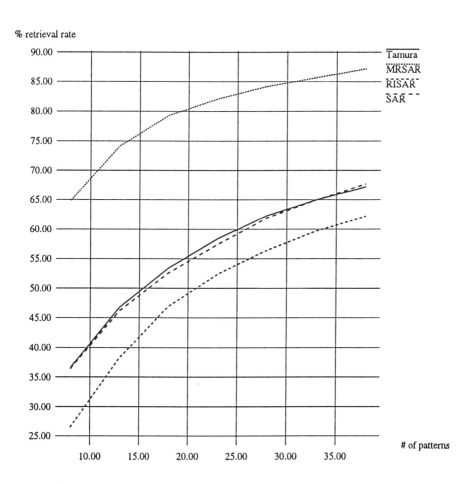

Figure 11.7 Evaluation of texture-based image retrieval.

sions of the original images. In the standard database, however, the MR-SAR works better (see Figure 11.7) since it uses more features per level (the RISAR uses only two model parameters, one for each circle).

Tamura Features on Brodatz The following factors should be taken into account while computing the coarseness and directionality measures on Brodatz images [KZL94]:

Algorithm	Time (sec)
SAR	1.6
RISAR	2.1
MRSAR	2.3
Tamura	9.9

Table 11.2   Texture algorithm timings.

- Resolution reduction must be chosen to provide both efficient computation and acceptable accuracy; reducing the resolution by a factor of 4×4 and computing for window sizes of $1 \times 1, \ldots, 8 \times 8$ seems to be suitable.

- The modification of the coarseness measure for small size regions is not required for the Brodatz images since they are fairly uniform.

- The algorithm for determining peaks in the directionality histogram must be modified to deal with multiple peaks (the original algorithm [TMY76] allows a maximum of only two peaks).

- Modification of the directionality measure to handle heterogeneous texture orientations performs better than the unmodified measure.

- Unlike the SAR features, the Tamura features are almost uncorrelated; as a result, the simplified Mahalanobis distance gives as good accuracy as the Mahalanobis distance.

Comparison of SAR and Tamura features   Figure 11.7 summarizes the results of evaluating models based on Tamura features, single-resolution SAR, RISAR, and four-resolution MRSAR. The time required for feature computation for a 128×128 image on a Sun SPARCstation 10 platform is shown in Table 11.2. Retrieval can be accomplished in real-time when image features are pre-computed. The results show that the Tamura and SAR methods are of the same order of accuracy (36–68%), although the Tamura method requires more computation. The four-resolution MRSAR model gives the best results by far (65-88%), since it incorporates multi-scale information. Its disadvantage is that it is not possible to compute lower resolutions of the image when the image region is too small to start with (SAR features cannot be computed reliably for sizes smaller than 16×16).

The Tamura measures, though computationally most expensive, have the advantage of being rotation-invariant (they perform better than the RISAR model (26–62%)). An additional advantage is that the Tamura features are thought to be closer to human visual perception. An example of the ten nearest images retrieved for one of the Brodatz images (D101) is shown in Figure 11.8. The query image is the left-most image in the top row of the texture picker. Out of the ten retrieved texture images, seven belong to the same class, while the rest are from another class (D102) which has the same texture with inverted gray-levels.

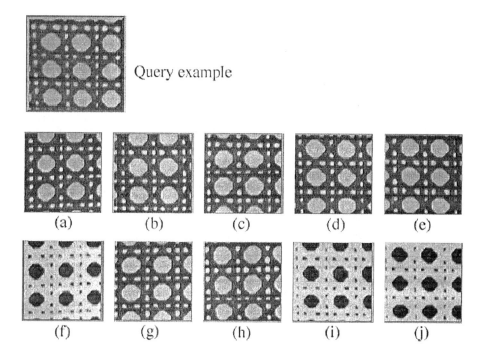

Figure 11.8 An example of retrieval of Brodatz textures based on Tamura features.

11.2.3 Object Shape

Color and texture are useful features in representing both scenes and the objects within them. However, there are also properties which may only be defined over individual objects. For purposes of indexing and retrieval, the most important of these features are geometric properties of objects, including shape, size, and location.

Suppose that segmentation decomposes an image into n objects. Let i be the index which identifies these objects, ranging from 1 to n. The *size* of the ith object, which we shall denote by S_i, will be defined simply to be the number of pixels in that particular segment. In the interest of economy of representation, it is usually a good idea to represent size in terms of a basic unit larger than a pixel, such as 1/36 of the entire image. In this case only those regions whose size is greater than the basic unit will be recognized as "objects." Once all of the pixels of the object have been identified, it is also possible to define their center of mass, (x_{ci}, y_{ci}) and define that point to be the *location* of the object.

The *shape* of an object is more difficult to define, since similarity of shape is often a matter of subjective interpretation. In image analysis the classical shape representation feature is a set of moment invariants . For simplicity, consider the case that object R is represented as a binary image; then a moment of order $p + q$ (where p and q are natural numbers) for the shape bounding object R is defined as follows:

$$m_{p,q} = \sum_{(x,y)\in R} x^p y^q \tag{11.20}$$

To make these moments invariant to translation, *central moments* are defined in terms of the object's center of mass as follows:

$$\mu_{p,q} = \sum_{(x,y)\in R} (x - x_{ci})^p (y - y_{ci})^q \tag{11.21}$$

The central moment can then be normalized for scale invariance [Jai89]:

$$\eta_{p,q} = \frac{\mu_{p,q}}{\mu_{0,0}^\gamma} \tag{11.22a}$$

$$\gamma = \frac{(p+q+2)}{2} \tag{11.22b}$$

Based on these moments Hu [Hu77] has developed a set of seven moment invariants with respect to translation, rotation and scale differences:

$$\phi_1 = \mu_{2,0} + \mu_{0,2} \tag{11.23a}$$

$$\phi_2 = (\mu_{2,0} - \mu_{0,2})^2 + 4\mu_{1,1}^2 \tag{11.23b}$$

$$\phi_3 = (\mu_{3,0} - 3\mu_{1,2})^2 + (\mu_{0,3} - 3\mu_{2,1})^2 \tag{11.23c}$$

$$\phi_4 = (\mu_{3,0} + \mu_{1,2})^2 + (\mu_{0,3} + \mu_{2,1})^2 \tag{11.23d}$$

$$\phi_5 = (\mu_{3,0} - 3\mu_{1,2})(\mu_{3,0} + \mu_{1,2})\phi_x$$
$$+ (\mu_{0,3} - 3\mu_{2,1})(\mu_{0,3} + \mu_{2,1})\phi_y \tag{11.23e}$$

$$\phi_6 = (\mu_{2,0} - \mu_{0,2})[(\mu_{3,0} + \mu_{1,2})^2 - (\mu_{0,3} + \mu_{2,1})^2]$$

$$+ 4\mu_{1,1}(\mu_{3,0} + \mu_{1,2})(\mu_{0,3} + \mu_{2,1}) \tag{11.23f}$$

$$\phi_7 = (3\mu_{2,1} - \mu_{0,3})(\mu_{3,0} + \mu_{1,2})\phi_x$$
$$+ (\mu_{3,0} - 3\mu_{2,1})(\mu_{0,3} + \mu_{2,1})\phi_y \tag{11.23g}$$

$$\phi_x = (\mu_{3,0} + \mu_{1,2})^2 - 3(\mu_{0,3} - 3\mu_{2,1})^2 \tag{11.23h}$$

$$\phi_y = (\mu_{0,3} + \mu_{2,1})^2 - 3(\mu_{3,0} + \mu_{1,2})^2 \tag{11.23i}$$

The other two objective measures which tend to capture salient information about shape are *circularity* and *major axis orientation* . A general approach in digital image processing [Pra91] is to define circularity in terms of size and perimeter, P_i. To a first approximation, perimeter may be defined as the number of pixels on the boundary; a more accurate approximation will take into account the difference between diagonal distance and horizontal and vertical distance. The formula for circularity, γ_i, is given as follows:

$$\gamma_i = \frac{4\pi S_i}{P_i^2} \tag{11.24}$$

This value ranges between 0 (which would correspond to a perfect line segment) and 1 (which would correspond to a perfect circle). Major axis orientation, θ, is defined in terms of the second-order central moments of the pixels in the region, $\mu_{1,1}$, $\mu_{2,0}$, and $\mu_{0,2}$, as follows [Jai89]:

$$\theta = \frac{1}{2} \arctan \frac{2\mu_{1,1}}{\mu_{2,0} - \mu_{0,2}} \tag{11.25}$$

The arctan function is usually interpreted to have a range between $-90°$ and $90°$. Other features which may be used to represent object shapes include elongation (the ratio of minimum and maximum distances from the center of mass to the boundary), number of holes and corners, and symmetry [Jai89]. Thus, as is the case with texture, shape is represented by a vector of feature values; and two such vectors may be compared for similarity using the same metrics outlined in Section 11.2.2.

11.2.4   Sketch-Based Features

One of the more intuitive approaches to describing an image in visual terms for the sake of retrieving it is to provide a sketch of what is to be retrieved. The richness of features in such a sketch will, of course, depend heavily on the skill of the user who happens to be doing the sketching, since a well-drawn sketch can make good use of techniques such as coloring and shading. However, as sort of a "lowest common denominator," we may assume that a sketch is a simple line drawing which captures some basic information about the shapes and orientations of at least some of the objects in the image.

Under this assumption the features of an image which would guide any attempt at retrieval would be some set of edges associated with a reasonably abstract representation of the image. One technique for constructing such an edge-based representation has been developed by the Electrotechnical Laboratory at MITI in Japan. This technique is a relatively straightforward five-step process [K+92]:

1. The image is normalized by an affine transformation to occupy a square area 64 pixels on each side; a median filter is then used to normalize the intensity values of the red, green, and blue color components.

2. For each pixel p_{ij} of the original image, a normalized gradient ∂_{ij} is computed for each color intensity value according to the following formula:

$$\partial_{ij} = |\Delta p_{ij}|/|I_{ij}| \tag{11.26}$$

$|\Delta p_{ij}|$ is the maximum of the differences between the pixel intensity and that of its immediate neighbors. $|I_{ij}|$ is the local power of the pixel intensity (where the power is determined by the number of immediate neighbors examined).

3. Let μ and σ be the mean and standard deviation, respectively, of all gradient values in the original image. Select those pixels p_{ij} for which:

$$\partial_{ij} \geq \mu + \sigma \tag{11.27}$$

These pixels will be called the *global edge candidates*.

4. For each global edge candidate, p_{ij}, let μ_{ij} and σ_{ij} be the mean and standard deviation of its gradient values to its immediate neighbors. Select those pixels for which:

$$\partial_{ij} \geq \mu_{ij} + \sigma_{ij} \tag{11.28}$$

These pixels will be called the *local edge candidates*.

5. Apply thinning and shrinking morphological operations [Jai89] to the local edge candidates to obtain an abstract edge representation.

It is worth observing that this technique is supported by physiological justification, since the contrast sensitivity of the human eye is proportional to the logarithm of the perceived intensity. The ratio computed in the second step is a digital representation of this proportionality.

In this case the measurement of similarity will be applied to an abstract image which is the result of this process, $P = \{p_{ij}\}$, where both i and j range between 1 and 64. That image will then be compared against a pixel representation of a sketch $Q = \{q_{ij}\}$, where i and j range over the same values. In this case comparison is a matter of both local and global correlation:

1. Divide both P and Q into 8×8 local blocks.

2. Apply a correlation computation ρ to the comparison of corresponding local blocks P_m and Q_m. However, also compute correlations between P_m and horizontal and vertical displacements of Q_m, where ϵ gives the extent of horizontal displacements and δ gives the extent of vertical displacement. Define the *local correlation* for block P_m to be C_m, the maximum of all computed correlation values.

3. Define the *global correlation* to be the sum of local correlations for all local blocks.

Note that this technique measures *similarity*, rather than *distance*; the higher the global correlation value, the better the match between the two images.

An example of experimental results using sketch-based indexing and retrieval is shown in Figure 11.9. In this example a sketch image was presented as a query, and the above sketch matching algorithm was applied in the search for the best matches. As one can see, the top five candidates agree very well with the query image, indicating the effectiveness of the technique. The major drawback of this straightforward sketch matching approach is that it is orientation and scale dependent. Similar sketch patterns with different orientation or scale will not be retrieved by this approach when compared with the query image. Overcoming this problem requires more sophisticated edge representation and matching algorithms.

11.3 INDEXING SCHEMES

A major problem in dealing with large image databases is efficiency of retrieval. One of the key issues in achieving such efficiency is the design of a suitable

Figure 11.9   An example of sketch-based image retrieval.

indexing scheme. Content-based image retrieval can only be effective if it entails reducing the search from a large and unmanageable number of images to a few that the user can quickly browse. If image features, such as those discussed in Section 11.2, are pre-computed before images are added to the database, then efficiency is best achieved if they are stored according to an index structure which contributes to pruning this search space.

11.3.1   Multi-Dimensional Indexing Methods

Generally speaking, an index in a database consists of a collection of entries, one for each data item, containing the value of a key attribute for that item and a reference pointer that allows immediate access to the item. If image features are pre-computed for each image in a database, then the key attribute for that image will be a *feature vector* which corresponds to a point in a multi-dimensional feature space; similarity queries will then correspond to nearest neighbor or range searches. Thus, fast query processing requires a multi-dimensional indexing technique which will function efficiently over a large database.

There are three popular approaches to multi-dimensional indexing: R-trees (particularly the R*-tree); linear quadtrees; and grid files. However, most of these multi-dimensional indexing methods explode geometrically as the number of dimensions increases; so for a sufficiently high dimensionality, the technique is no better than sequential scanning [F+94]. For example, search time with linear quadtrees is proportional to the hypersurface of the query region; and the hypersurface grows geometrically with the dimensionality. With grid files it is the directory that grows geometrically with the dimensionality. Methods based on R-trees, on the other hand, can be efficient if the fan-out of the R-tree nodes remains greater than 2 and the dimensionality stays under 20.

The limitations of these indexing methods may be attributed to two underlying assumptions:

1. Distance in the search space is Euclidean.

2. The dimensionality of the search space is reasonably low.

It has already been demonstrated in Section 11.2.2 that Euclidean distance is seldom an effective measure of proximity. Furthermore, if an image is going to be indexed by a viable combination of the features summarized in Section 11.2, then the dimensionality of the feature vector will easily exceed what these techniques assume to be "reasonably low."

This problem is best solved by appealing to the philosophy articulated in Section 11.1: The search space should be *filtered* by a preprocessing step which may allow some false hits, but not false dismissals. Mathematically, this preprocessing requires mapping each feature vector $\overrightarrow{X}$ into a vector $\overrightarrow{X'}$ in a more suitable space, in which all distances will underestimate the distances in the actual search space [F+94]. This mapping will guarantee that a query based on $\overrightarrow{X'}$ space will not miss any actual hit, but may contain some false hits. This reduces the problem to defining a suitable $\overrightarrow{X'}$ space. Three approaches to solution will now be discussed: reducing dimensionality of the feature space with coarse representation and grouping, constructing an indexing tree based on a classification derived from a neural network, and interactive text annotation.

11.3.2 Coarse Representation and Grouping

There are a variety of approaches to the coarse representation or grouping of image features, each of which will be reviewed.

Incremental Histogram Intersection

As was described in Section 11.2.1, the first step towards efficient histogram comparison is the reduction of the number of bins to be compared. Efficiency may be further improved with a so-called *incremental* approach to intersection computation [SB91]. Returning to the notation of Section 11.2.1, let I be an image in the database; and let I_i be its most populated bins, sorted in descending order of size. (In the experiments reported in [SB91], i ranged from 1 to 10.) Each of these bins corresponds to a specific balance of color components, called its *index*; and it can be assigned a *key* which is the fraction of the total number of pixels in the image that it contains. Now suppose Q is a query image to be compared for similarity to the images in the database. Construct Q_i by the same process, again assigning to each bin a key which is a fraction of the total number of pixels. The problem is to find those images I which have high histogram intersection with Q. Begin the search with Q_1. Let C_1 be the index of this bin. Search for those images I which have a bin of index C_1 whose fractional key value is greater than or equal to that of Q_1. Iteration may then be used to filter this initial set by accounting for similar representation of the indexes corresponding to the remaining Q_i.

The complexity of this process is $O(N \log N + cM)$, where M is the total number of images in the database, N is the total number of histogram bins, and c is the number of bins in the reduced representation [SB91]. On the other hand the complexity of simple intersection, as described in Section 11.2.1, is linear with respect to the product of the number of histogram bins and the size of the database, $O(NM)$. However, a large-scale database requires further reduction in the complexity of the searching process. This may be achieved by combining the incremental technique with a coarser representation of the histogram in terms of numerical keys. This approach will now be reviewed.

Coarse Histogram Representation

The basic approach to representing a histogram as a small number of numerical keys is a geometrical one [G+94]. If we represent each histogram bin as a point in its associated color space, then a subset of n representative bins forms a hyper-polygon if all these points are connected in a certain order. Such a hyper-polygon may be represented by two parameters, perimeter (P) and angle (A), which may be defined as follows:

$$P = \sum_{i=1}^{n} \frac{(1+c_i)}{2} d_{i-1,i} \tag{11.29a}$$

$$A = \sum_{i=1}^{n} \frac{(1+c_i)}{2} \alpha_i \tag{11.29b}$$

Both of these definitions are weighted sums, weighted by $(1 + c_i)/2$ where c_i is the size of bin i. $d_{i-1,i}$ is the distance between two successive points in the color space corresponding to bins i and $i-1$. (The 0th point is the origin in the space.) Finally, α_i is the angle between the two lines which meet at the point corresponding to bin i (the nth point being connected back to the origin). This reduces the entire histogram image to only two real number parameters.

The effectiveness of this representation depends on the uniqueness of each hyper-polygon of a database image. This uniqueness depends, in turn, on the order in which the bins are sorted prior to calculating P and A. The simplest approach is to sort the bins by their pixel counts. However, this approach may lead to false negatives during search: similar images which are likely to be rejected from the candidate list. This is because even though the representative histogram of a database image may have exactly the same bins as the query specification, the size of some of the bins may be different; and this may yield very different P and A values from those of the query image. To avoid this problem, the histogram bins are sorted in ascending order by their distance from the origin in the color space. Figure 11.10 depicts such sorting of five histogram bins, with the origin linked to the first and last of these bins. The index parameters P and A are calculated based on the polygon defined by such linking.

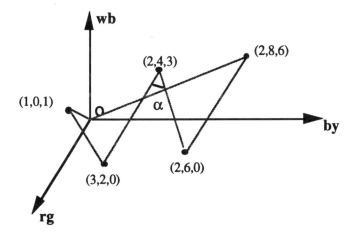

Figure 11.10 Sorting histogram bins for calculating P and A in opponent color space.

Group number	Size range
1	$\frac{1}{4}S_A < S \leq \frac{1}{2}S_A$
2	$\frac{1}{2}S_A < S \leq S_A$
3	$S_A < S \leq 2S_A$
4	$2S_A < S \leq 3S_A$
5	$3S_A < S \leq 4S_A$
6	$4S_A < S \leq 5S_A$
7	$5S_A < S \leq 6S_A$
8	$6S_A < S \leq 7S_A$
9	$7S_A < S \leq 8S_A$
10	$8S_A < S \leq 9S_A$

Table 11.3   Groups of object sizes.

Coarse Representation of Other Object Features

In the representation of texture, there has been research in applying steerable filters [FA92] over multiple scales; but this approach can only be used for the representation of directionality [PG93]. More general approaches to the coarse representation of texture have not yet been explored. On the other hand shape, size, location , and color may all be coarsely grouped as follows:

Location: In Section 11.2.1 we discussed partitioning an image into a 3 × 3 array of 9 sub-areas. These sub-areas may also be used for a coarse representation of location. The location of an object is just the label (usually a number from 1 to 9) of the sub-area which contains its center of mass, no matter how large the region size is.

Size: The coarse representation of the size of an object, denoted by S, is based on the size S_A of the sub-areas into which the entire image is partitioned; the pixel count is translated into a group number as defined in Table 11.3.

Shape: Since circularity ranges between 0 and 1, it can be simply divided into four groups, each with a range of 0.25. Similarly, orientation can be divided into 45° segments. Each of these groups can be labeled by a sequential numbering. Therefore, object shape can be represented by a total of 16 groups.

Color: The HVC color space can be similarly divided into seven sectors, each of which is clearly distinguishable for the others by subjective perception. One of these sectors is an achromatic zone with Chroma ≤ 20. The remaining six sectors are distinguished only by intervals in the Hue range.

The index keys presented above may be too coarse for a very large database, and the size of a retrieval result may be too large for user browsing. Therefore, as was discussed in Section 11.3.1, they will only be used as a high-level index to narrow down the search space; and a second level of higher-resolution indexing is necessary. The finer level provides the original representations of all index features. Thus, if the number of images retrieved by the coarse level of search exceeds some threshold, a finer matching based on techniques such as incremental histogram intersection may be performed on the results of that coarse search.

11.3.3 Indexing by Abstraction and Classification

The Iconic Index

As was discussed in Section 11.3.1, most databases use tree indexing for efficiency. An effective tree may be constructed either top-down, using classification, or bottom-up, using abstraction. Indexing by classification or abstraction is especially useful and feasible when the nature of the objects (such as faces) in the database are known *a priori*. This is the model-based approach briefly discussed in Section 11.1; and the structure of the model may be exploited to structure the index. For example, suppose a model represents every object o_i as a set of attribute-value pairs: $o_i\{(A_1, a_1), (A_2, a_2), \ldots, (A_n, a_n)\}$, where a_j is the value of the attribute A_j. Then, a tree index can be constructed by the following steps [Cha89]:

1. **Labeling:** For a given attribute-value pair (A_j, a_j), identify and label all o_i whose A_j value is either identical to a_j or lies with a certain range of a_j.

2. **Clustering:** Cluster together all objects which have the same label to form an abstraction of the (A_j, a_j) attribute-value pair.

3. **Indexing:** The abstraction created at Step 2 can then be represented as a node in the index tree. Steps 1 and 2 may now be repeated for another value of A_j. After all possible abstracted objects have been created for A_j, the nodes at that level may be similarly abstracted to construct the next level of the index tree.

This recursive process, known as *abstraction*, will ultimately result in a single cluster whose node corresponds to the index tree for the A_j attribute. A reverse process, known as *classification* , proceeds in a similar fashion, but by *dividing* the set of all objects, rather than clustering individual objects [Cha89].

The biggest problem with this approach is that it is rarely desirable to construct an index tree with respect to a single attribute. However, a similar technique may be applied in a multi-dimensional attribute-value space. In this case Steps 1 and 2 may be collapsed into a single Step, consisting in identifying a specific point in this space and then collecting all objects which lie within a certain distance from that point according to the similarity metric for that space. Another approach is to cluster the nodes of the tree according to different attributes at each level, but this approach assigns a priority to the different attributes which will impact the subsequent search strategy. If such a priority cannot be realistically defined, then the multi-dimensional approach is likely to be more successful.

If the attribute-value model is grounded in the semantics of the images being indexed, then the labels assigned to the index nodes may be based on that semantic interpretation [W+94]. More importantly, that semantic interpretation may be applied to represent each node in the tree by a suitably constructed iconic image . A straightforward way to construct such an icon is to select that image which is closest to the mean or centroid of the set of images corresponding to the node. This iconic representation of the index provides a useful browsing tool for quickly viewing representative images of the index categories. Such browsing can be further facilitated by allowing horizontal and vertical zooming in the viewing of the iconic index trees. [W+94] provides a good example of such iconic index trees for a database of faces.

Constructing an Iconic Index with a Self-Organizing Map

The Self-Organizing Map (SOM) [Koh90] uses a technique known as *linear autoassociative learning* to classify a set of input data with a neural network. Not

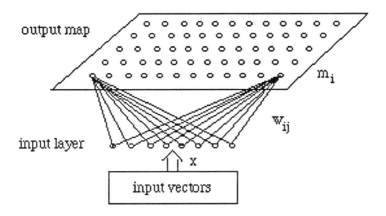

Figure 11.11 SOM architecture.

only is it well-suited to constructing index trees, but the resulting categories frequently admit of semantic interpretations which enable the construction of useful image icons. Figure 11.11 is an illustration of a typical two-layer SOM network. The input consists of vectors from an n-dimensional feature space R^n. The output (known as the *map*) is a regular two-dimensional grid. The number of nodes, M, in the map is more than the number of possible classification categories. Each node j in the input layer (corresponding to one of the components of the feature vector input) is connected to each node i in the map grid with an associated weight w_{ij}. All the weights associated with node i in the map grid are defined as the *reference vector* of this node; since there are n of these weights, the reference vector is n-dimensional, $m_i \in R^n$.

Every SOM is characterized by its set of weights w_{ij}. These weight values are initially assigned random values and are then set by the autoassociative learning process as follows:

■ Select a feature vector f for input.

■ Find the map node, i, whose reference vector is closest to the f according to a proximity metric, such as Euclidean distance. This node is called the *winning node*.

■ Adjust the reference vector of the winning node and its neighbors using
   the *delta rule* (also known as the "Widrow-Hoff learning rule") [RHM86].

This process iterates until the adjustments all approach zero, the criterion for
convergence at which time learning terminates. The resulting set of weights
should then map each input vector to a grid node whose reference vector is
closest to it in the n-dimensional feature space. The grid nodes can then be
clustered into classes [Koh90] corresponding to nodes of a level in an index
tree. Applying the same process to the representative feature vectors of each of
these nodes will then create the next level of the tree, and the process may be
repeated until the root of the tree is reached. This technique has been applied
to a variety of classification tasks, but a specific example of the construction of
an image index is given in [ZZ95].

It is also possible to take a less random approach to the learning process [W+94].
In this case the training set includes not only a set of input feature vectors but
also a template vector for each grid node. This method is very effective for
classifying objects for which a number of classes have been pre-defined, since
"knowledge" of those classes is embodied in the template vectors. Thus, the
technique has been successfully applied to constructing iconic indexes for facial
data.

11.3.4   Automatic Annotation Using Image Features

As was discussed in Section 11.1, the conventional method for indexing image
databases is to associate key words with each image. While the whole premise
of this Chapter is that there are several problems inherent in systems which
are exclusively text-based, key words can definitely complement visual content
retrieval, not to mention the fact that a great population of database users
are so used to text-based retrieval that giving up key words would be a great
inconvenience. Thus, there remains a need for automated tools to assist, if not
replace, human operators, who are costly, subjective, and sometimes unreliable,
in generating key words. It is therefore worth reviewing some current research
ideas in developing tools for annotation based on image features, particularly
textures and color histograms.

Though fully general automatic generation of key words is still beyond the
current technology of computer vision and is unlikely to be realized in the

near future, with human input and feedback, a large part of the annotation of image data may be generated from less "semantic" features of the images. Assume that a user labels a piece of an image as "water;" then a feature representation or model of that piece of image can be used to propagate this label to other "visually similar" regions. This approach has been developed into a system which assists a user in annotating large sets of images based on texture properties of those images [PM94].

In this system all images in a database are preprocessed to extract their features based on a set of texture models. The image is segmented into patches each of which is represented by a vector of those features, and a clustering process is applied to the feature vectors to generate an index tree. A user can then browse through the images in the database and select patches from one or more images as "positive examples" for an annotation. Based on the selection of patches, the system selects the "best" model(s) to represent the annotation label and propagates the label to other regions of images in the database that should have the same label based on the model(s). Falsely labeled patches can be removed by selecting them to be "negative examples" after viewing the first labeling result; and the system remembers each of the positive and negative examples selected by the user by updating the labels accordingly. An annotation of the entire database is thus generated through interaction between the user and the system.

A key asset of this approach is the use of multiple texture models, since no single feature model has been found to be good enough to reliably match human perception of similarity in pictures. Rather than using one model, the system is equipped with the ability to choose the one which "best explains" the regions which the user has selected as being significant for annotation. If none of the available models suffices, the system can combine models to create better explanations. An experimental test based on 98 images of natural scenes showed that the system could acquire an average gain of four to one in label prediction [PM94].

Figure 11.12 shows a similar example in which the color histogram model defined in Section 11.2.1 is applied to labeling the road in a natural scene. Here, A, B, and C are three patches that the user selected as "positive examples" of "road;" and the other patches outlined are similarly identified by the system using the color histogram model. Even without any fine tuning of the model, the selected regions agree very well with the examples.

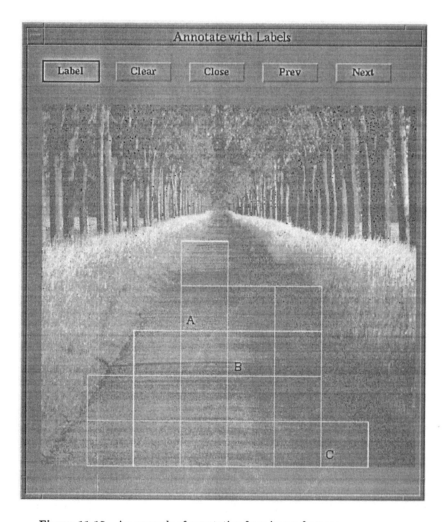

Figure 11.12 An example of annotation from image features.

11.4 A USER INTERFACE FOR FORMULATING QUERIES AND BROWSING RESULTS

As was observed in Section 11.1, a user-friendly environment for both flexible formation of queries and browsing results is crucial for image databases supporting visual content-based retrieval. Queries are most likely to be effective if they can be specified visually, interactively, and iteratively; and the same is true for assessing retrieval results. This is understandable since visual information can be recognized and understood faster than text descriptions, resulting in retrieving more relevant images. Also, only with visual feedback can we exploit the unique characteristics of retrieved images to refine a query for subsequent retrievals.

11.4.1 Query Formation

A query may be formulated by either some form of painting or sketch by the user or by selection from example images. Examples may be used as sources of color and texture patterns, sketches, layout or structural descriptions, or other graphical information. To accommodate these options in an interface, there are three basic approaches: template manipulation, object feature specification, and visual examples.

Query by Visual Templates

Visual templates may be used when a user wants to retrieve images which consist of some known color and/or texture patterns, such as a sunny sky, sea, beach, lawn, or forest. To support this, a set of pre-defined templates are stored and can be assigned to *template maps*. A template map is based on the same division of an image into a 3 × 3 array of 9 sub-areas which was discussed in Section 11.2.1. Different templates may be assigned to any subset of these sub-areas, and the resulting assignment serves as a query image. This approach is especially important if the user cannot provide any sort of painting or sketch and cannot find a useful example image. The map also has the advantage that unassigned areas serve as "don't care" specifications. Thus, Figure 11.13 illustrates an example of a query which is formulated strictly with respect to the color content of the left and right periphery. As a result, the retrieved images exhibit considerable variation in their central areas.

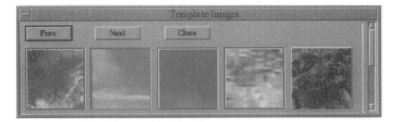

Figure 11.13   A menu of sample images for query by visual templates.

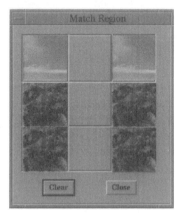

Figure 11.13 *(continued)*
A query image composed from selections from the menu.

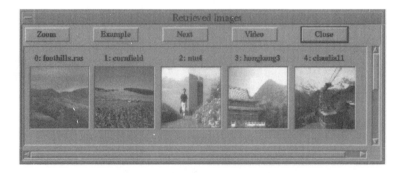

Figure 11.13 *(continued)*
The result of retrieval from the composed query.

This environment provides a list of templates which can be selected either visually from their icons or from their names. A user can further manipulate the color distribution of a selected template by adjusting three scalars corresponding to three color components of up to a certain number of the most dominant colors of which the selected template is composed. The user can also define new templates, which may be used along with the pre-defined templates.

Drawing a Query

A natural way to specify a visual query is to let the user paint or sketch an image. A feature-based query can then be formed by extracting the visual features from the painted image. The sketch image shown in Figure 11.9 is an good example of such a query: the user draws a sketch with a pen-based device or photo manipulation tool. The query may also be formed by specifying objects in target images with approximate location, size, shape, color and texture on a drawing area using a variety of tools, similar to the template maps, to support such paintings. This is one of the interface approaches currently being pursued in QBIC [N+93]. In most of the cases, coarse specification is sufficient, since a query can be refined based on retrieval results. This means it is not necessary for the user to be a highly skilled artist. Also, a coarse specification of object features can absorb many errors between user specifications and the actual target images.

Query by Visual Examples

A user may still have difficulty specifying an effective query, perhaps for want of a clear idea of what is desired. Such a user may formulate an incomplete query which retrieves a preliminary set of images. However, it may be that none of these are correct; but there is one which is visually similar to the desired target. In such cases the user can select that image as an example, and the system will search for all images which consist of similar features. To support such query by example , two options should be provided in the interface: a query may be based on either the entire example image or a specific region of that image.

Incomplete Queries

Supporting incomplete queries is important for image databases, since user descriptions often tend to be incomplete. Indeed, asking for a complete query is often impractical. To accommodate such queries, search should be confined to a restricted set of features (a single feature value, if necessary). All query formation environments presented above can provide the option of specifying which features will be engaged in the search process. For instance, a query can be formed based on template selection that will retrieve all images containing a red car, regardless of whether the car is in the center or any other part of the image and whatever size it may be.

11.4.2   Browsing and Feedback for Image Selection and Query Refinement

As illustrated in Figure 11.1, the query and retrieval environment also must support efficient browsing of images and feedback from retrieval; so query responses can be viewed and further refined. Browsing can be especially helpful when the specification of pictorial content is ambiguous or incomplete, and it may be the only method for making the final selection. A good method for supporting this is to present results from a query as an array of thumbnail images (100×100 pixels) in a browsing window, as shown in Figure 11.9 and Figure 11.13, thus facilitating quick review. A zooming function can be provided to allow users to examine any image in up to its original pixel resolution. In the browsing window any image can also be selected as an example image to launch a new query; this is enabled with the "Example" button in Figure 11.13. Finally, such a system should also support query refinements using a combination of images (an entire image or part of an image). Thus, the user should be able to formulate a query such as: "find me all images which contain object X as in image A but has the same background as in image B."

11.5   PERFORMANCE EVALUATION

The usual metrics for evaluating the performance of a retrieval system are *recall* and *precision* [SM83]. Recall measures the ability of the system to retrieve relevant items, while precision conversely measures the ability to reject irrelevant items. For information systems based on exact match, these measures are defined as the following ratios:

- *Recall* is the ratio of the *number of items retrieved and relevant* to the *total number of relevant items in the collection.*

- *Precision* is the ratio of the *number of items retrieved and relevant* to the *total number of items retrieved.*

These two measures are interdependent, and usually one cannot be improved without sacrificing the other. If recall cannot be improved without sacrificing precision (or vice versa), then, ultimately, it may be a matter of user preference as to what constitutes optimal performance. That is, the user must determine which metric is preferable and how much the other can be sacrificed. In typical retrieval systems recall tends to increase as the number of retrieved items increases; at the same time, however, the precision is likely to decrease.

When we apply these two performance measures to evaluating image databases, we have to take into account the fact that similarity ranking, as opposed to exact match, is used in the retrieval process. In this case, *normalized* recall and precision measures [SM83] can be used to reflect the positions in which the set of relevant items appear in the retrieved image list. Thus, normalized recall measures how close to the top of the list of retrieved images the set of similar images appears, compared to the ideal retrieval in which all similar images appear as the top candidates. The most comprehensive study of using the recall and precision measures to assess the performance of image feature based retrieval may be found in [F+94].

11.6 IMAGE RETRIEVAL IN PRACTICE

11.6.1 Areas of Application

As was observed in Section 11.1, there are a wide variety of applications which can take advantage of databases which accommodate images as well as text. An examination of several key examples of these applications has been provided by [ANAH93]. Let us now review the significant elements of these examples.

Multimedia Documents

The print medium has never been restricted to text. This very book could not possibly communicate as effectively as it does without the benefit of the images provided in its figures. Often when we read such documents, the memories of the images we see may be more salient than those of the text we read, yet those images can only be indexed by the text in their captions . . .if they are indexed at all. While the ability to find the location of an image with certain visual properties may not be particularly critical for this book, the problem becomes

more relevant if the book is about art history or a highly "visual" science, such
as biology or geology.

Transaction Systems

More and more merchandising is being conducted through catalog sales. Such
catalogs, again, are rarely effective if they are restricted to text. The customer
who browses such a catalog is more likely to retain visual memories than text
memories, and the catalog is frequently designed to cultivate those memories.
However, if browsing whets the taste for buying, finding the product one wishes
to order may be impeded if text is the only instrument for retrieval. If the
catalog is small enough, one can, of course, simply browse it again; but, as the
catalog grows in size, the problem of recovering that particular image which
inspired the need to purchase becomes more formidable. If that image cannot
be conveniently recovered, that urge to buy may gradually dissipate.

Trademark Systems

Section 11.2 presented a systematic review of those properties which are involved when we characterize two images as similar or different. Ultimately, it
addresses the question of how *qualitative* differences may be expressed *quantitatively*. Such quantification is particularly important in the development of
appropriately efficient indexing schemes, such as those discussed in Section 11.3.
However, this quantification of difference of appearance may serve other needs
as well. For example a trademark may only be assigned if it does not resemble
any other trademarks too closely. Traditionally, such proximity of resemblance
has been a matter of *subjective* judgment. If such decisions can be translated
into quantitative terms, one is more likely to be able to establish an *objective*
case for the uniqueness of a proposed new trademark.

Medical Applications

Biology and geology were cited above as being "visual" sciences. Medicine is
an area in which visual recognition is often a significant technique for diagnosis.
The medical literature abounds with "atlases," volumes of photographs which
depict normal and pathological conditions in different parts of the body, viewed
by the unassisted eye, through a microscope, or through various imaging technologies. An effective diagnosis may often require the ability to recall that a
given condition resembles an image in one of these atlases. The atlas allows the

Querying	Browsing
Mug shot identification	Multimedia encyclopedia
Museum inventory	Medical encyclopedia
Trademark retrieval	Photographic banks
Medical atlases	Sales catalogs
Geographical information systems	

Table 11.4 Typical applications for querying and browsing.

reader to interpret that image in terms of patient state, and that interpretation can then lead to diagnosis [Hun89].

Needless to say, the amount of material cataloged in such atlases is already large and continues to grow. Furthermore, it is often very difficult to index with textual descriptions. As was observed in Section 11.1, a community of specialists often develops their own, highly specific vocabulary; but that vocabulary is only as good as the scope of conditions it describes. As soon as an image exhibits a property beyond that scope, the specialist is no better off than a novice in trying to retrieve the most appropriate image. If the most important element in retrieving an image from an atlas is resemblance of appearance, then, ultimately, retrieval of atlas images must be based on resemblance of appearance, rather than any text descriptions assigned to those appearances.

Criminal Identification

This is the area which was discussed in Section 11.1 as a good example of model-based image retrieval [W+94]. Victims and witnesses must often endure the tedium of examining large quantities of mug shots in the hope that one of them will match the suspect they saw. This material is often not organized in any systematic way, nor can it be effectively indexed. Often searching can only be guided by an incomplete query, such as those discussed in Section 11.4.1. However, if such a query may be formulated in visual terms, it is more likely to lead to effective identification than if one can only iterate through a very long list of candidates, most of which are too visually remote to be even viable.

Querying and Browsing

Most of the applications cited above require the processing of highly focused queries. However, casual browsing can often be as important a use of image resources as is query processing. Table 11.4 summarizes several typical applications on the basis of whether they are likely to support focused querying or casual browsing. This classification is not intended to be absolute, but it should provide a general sense of the current scope of applications for image retrieval.

11.6.2   Practical Implementations

While there is now extensive experimental work in the development of systems which support content-based image retrieval, the only viable product to have been released thus far has been the Ultimedia Manager 1.0 from IBM [Sey94]. This system runs on OS/2 and uses the DB2 database manager. It integrates text annotations of images with queries based on all the features discussed in Section 11.2. Future releases are already being planned which will add more imaging functions, support for more database managers (SQL support is a major target here), more platforms (including Windows, Macintosh, Solaris, and HPUX), and a client-server edition.

12

CONTENT-BASED VIDEO INDEXING AND RETRIEVAL

12.1 INTRODUCTION

Video will only become an effective part of everyday computing environments when we can use it with the same facility that we currently use text. Computer literacy today entails the ability to set our ideas down spontaneously with a word processor, perhaps while examining other text documents to develop those ideas and even using editing operations to transfer some of that text into our own compositions. Similar composition using video remains far in the future, even though work-stations now come equipped with built-in video cameras and microphones, not to mention ports for connecting our increasingly popular hand-held video cameras.

Why is this move to communication incorporating video still beyond our grasp? The problem is that video technology has developed thus far as a technology of images. Little has been done to help us use those images effectively. Thus, we can buy a camera that "knows" all about how to focus itself properly and even how to compensate for the fact that we can rarely hold it steady without a tripod; but no camera knows "where the action is" during a football game or even a press conference. A camera shot can give us a clear image of the ball going through the goal posts, but only if we find the ball for it.

The point is that we do not use images just because they are steady or clearly focused. We use them for their content. If we wish to compose with images as we currently compose with words, we must focus our attention on content. Video composition should not entail thinking about image "bits" (pixels), any more than text composition requires thinking about ASCII character codes. Video content objects include footballs, quarterbacks, and goal posts. Unfortunately, state-of-the-art software for manipulating video does not "know" about such objects. At best, it "knows" about time codes, individual frames, and clips of video and sound. To compose a video document—or even just incorporate video as part of a text document—we find ourselves thinking one way (with ideas) when we are working with text and another (with pixels) when we are working with video. The pieces do not fit together effectively, and video suffers for it.

Similarly, if we wish to incorporate other text material in a document, word processing offers a powerful repertoire of techniques for finding what we want. In video about the only technique we have is our own memory coupled with some intuition about how to use fast forward and fast reverse buttons while viewing. Thus, the effective use of video is still beyond our grasp because the effective use of its content is still beyond our grasp. In this Chapter we shall address four areas in which software can make the objects of video content more accessible:

Partitioning: We must begin by identifying the elemental index units for video content. In the case of text, these units are words and phrases, the entities we find in the index of any book. For video we speak of *generic clips* which basically correspond to individual camera shots (with certain exceptions which will be discussed in the sequel).

Representation and classification: Once a generic clip has been identified, it is necessary to represent its content. This representation may involve the use of text, mathematical transforms, or images. Representation also assumes that we have an *ontology* which embodies our objects of interest [LG89] and that video content may be classified according to this ontology. It is important to recognize that any given video may be classified according to multiple ontologies. Even a domain as simple as football may, on the one hand, be viewed with respect to an ontology of offensive and defensive strategies; but a sociologist may be more interested in examining it in terms of an ontology of body language.

Indexing and retrieval: One way to make video content more accessible is to store it in a database [SSJ93]. Thus, there are also problems concerned with how such a database should be organized, particularly if its records are to include images as well as text. Having established how material can be put *into* a database, we must also address the question of how that same material can be effectively *retrieved* , either through directed queries which must account for both image and text content or through browsing when the user may not have a particularly focused goal in mind.[1]

Interactive tools: The above three areas are all concerned with the *functionality* of software which manages video data. However, if accessibility is a key *desideratum*, then the *user interfaces* to that functionality will be as important as the functionality itself. We are not used to interacting with video; most of our experiences involve simply sitting and watching it passively. If video is to join the ranks of the information resources we utilize, then we shall need tools which facilitate and encourage our interacting with it. These tools will obviously take advantage of the functionality of the other three areas in this list, but they add value to that functionality by making it more likely that such functionality will actually be employed.

Each of these areas will now be discussed.

12.2 PARTITIONING TECHNIQUES FOR FULL-MOTION VIDEO

12.2.1 The Nature of the Problem

We begin with the assumption that a generic clip, our basic indexing unit, is a single uninterrupted camera shot. The partitioning task is then a problem of detecting boundaries between consecutive camera shots. The simplest transition is a *camera break*. Figure 12.1 illustrates a sequence of four consecutive video frames with a camera break occurring between the second the third frames. The significant qualitative difference in content is readily apparent. If that difference can be expressed as a suitable metric, then a segment boundary can be declared whenever that metric exceeds a given threshold. Hence, establishing such metrics and techniques for applying them is the first step towards automating the partitioning of full-motion video.

A camera break is the simplest transition between two shots. More sophisticated techniques include dissolve, wipe, fade-in, and fade-out [BT93]. Such special effects involve much more gradual changes between consecutive frames than does a camera break. Figure 12.2 shows five frames of a dissolve from

---

[1] As Martin Heidegger put it: "Questioning can come about as 'just asking around' or as an explicitly formulated question" [Hei77].

Figure 12.1   Four frames across a camera break from a documentary video:
The first two frames are in the first camera shot, and the third and fourth
frames belong to the second camera shot. There are significant content changes
between the second the third frames.

a documentary video: the frame just before the dissolve begins, three frames
within the dissolve, and the frame immediately after the dissolve. This sequence illustrates the gradual change that diminishes the power of a simple
difference metric and a single threshold for camera break detection. Indeed,
most changes are even more gradual, since dissolves usually last more than ten
frames. Furthermore, the changes introduced by camera movement, such as
pan and zoom, may be of the same order as that introduced by such gradual
transitions. This further complicates the detection of the boundaries of camera shots, since the artifacts of camera movements must be distinguished from
those of gradual shot transitions .

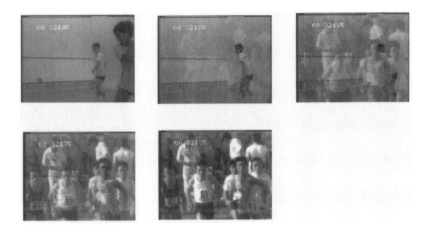

Figure 12.2   Frames in a dissolve: The first frame is the one just before the dissolve starts, and the last one is the frame immediately after the end of the dissolve. The rest are the frames within the dissolve.

We shall now present several difference metrics which have been applied for detecting simple camera breaks. We then discuss how the use of any of these metrics can be adapted to accommodate gradual transitions. In Section 12.3.1 we address how motion analysis may be applied to eliminating false detection of transitions resulting from the artifacts of camera movements.

12.2.2   Difference Metrics

Pair-wise Pixel Comparison

A simple way to detect a qualitative change between a pair of images is to compare the corresponding pixels in the two frames to determine how many pixels have changed. This approach is known as *pair-wise comparison* . In the simplest case of monochromatic images, a pixel is judged as changed if the difference between its intensity values in the two frames exceeds a given threshold t. This metric can be represented as a binary function $DP_i(k, l)$ over the domain of two-dimensional coordinates of pixels, (k, l), where the subscript i denotes the index of the frame being compared with its successor. If $P_i(k, l)$ denotes the intensity value of the pixel at coordinates (k, l) in frame i, then

$DP_i(k, l)$ may be defined as follows:

$$DP_i(k, l) = \begin{cases} 1 & \text{if } |P_i(k, l) - P_{i+1}(k, l)| > t \\ 0 & \text{otherwise} \end{cases} \qquad (12.1)$$

The pair-wise segmentation algorithm simply counts the number of pixels changed from one frame to the next according to this metric. A segment boundary is declared if more than a given percentage of the total number of pixels (given as a threshold T_b) have changed. Since the total number of pixels in a frame of dimensions M by N is $M \times N$, this condition may be represented by the following inequality:

$$\frac{\sum_{k,l=1}^{M,N} DP_i(k, l)}{M \times N} \times 100\% > T_b \qquad (12.2)$$

A potential problem with this metric is its sensitivity to camera and object movement. For instance, in the case of camera panning, a large number of objects will move in the same direction across successive frames; this means that a large number of pixels will be judged as changed even if the pan entails a shift of only a few pixels. This effect may be reduced by the use of a smoothing filter: before comparison each pixel in a frame is replaced with the mean value of its nearest neighbors. (A 3×3 window centered on the pixel being "smoothed" can be used for this purpose [ZKS93].) This also filters out some noise in the input images.

Another approach to robustness is to compare corresponding *regions* (blocks) in two successive frames on the basis of second-order statistical characteristics of their intensity values. One such metric for comparing corresponding regions is called the *likelihood ratio* [KJ91]. Let μ_i and μ_{i+1} denote the mean intensity values for a given region in two consecutive frames, and let S_i and S_{i+1} denote the corresponding variances. The following formula computes the likelihood ratio and determines whether or not it exceeds a given threshold t:

$$\frac{[\frac{S_i + S_{i+1}}{2} + (\frac{\mu_i - \mu_{i+1}}{2})^2]^2}{S_i \times S_{i+1}} > t \qquad (12.3)$$

Camera breaks can now be detected by first partitioning the frame into a set of sample areas. Then a camera break can be declared whenever the total number of sample areas whose likelihood ratio exceeds the threshold is sufficiently large (where "sufficiently large" will depend on how the frame is partitioned). An advantage that sample areas have over individual pixels is that the likelihood ratio raises the level of tolerance to slow and small object motion from frame to frame. This increased tolerance makes it less likely that effects such as slow motion will mistakenly be interpreted as camera breaks.

The likelihood ratio also has a broader dynamic range than does the percentage used in pair-wise comparison. This broader range makes it easier to choose a suitable threshold value t for distinguishing changed from unchanged sample areas. A potential problem with the likelihood ratio is that if two sample areas to be compared have the same mean and variance, but completely different probability density functions, no change will be detected. Fortunately, such a situation is very unlikely.

Histogram Comparison

An alternative to comparing corresponding pixels or regions in successive frames is to compare some feature of the entire image. One such feature that can be used for partitioning purposes is a histogram of intensity levels. The principle behind this approach is that two frames having an unchanging background and unchanging objects will show little difference in their respective histograms. The histogram comparison algorithm should be less sensitive to object motion than the pair-wise pixel comparison algorithm, since it ignores the spatial changes in a frame. One could argue that there may be cases in which two images have similar histograms but completely different content. However, the probability of such an event is sufficiently low that, in practice, such errors are tolerable.

Let $H_i(j)$ denote the histogram value for the ith frame, where j is one of the G possible gray levels. (The number of histogram bins can be chosen on the basis of the available gray-level resolution and the desired computation time.) Then the difference between the ith frame and its successor will be given by the following formula:

$$SD_i = \sum_{j=1}^{G} |H_i(j) - H_{i+1}(j)| \qquad (12.4)$$

If the overall difference SD_i is larger than a given threshold T_b, a segment boundary is declared. To select a suitable threshold, SD_i can be normalized by dividing it by the product of G and $M \times N$, the number of pixels in the frame.

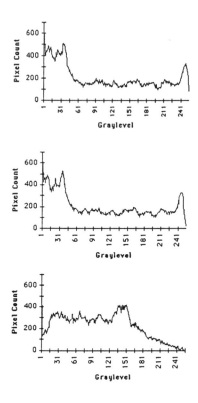

Figure 12.3 Histograms of gray-level pixel values corresponding to three successive frames with a camera break between the second and third frames.

Figure 12.3 shows gray-level histograms of the first three images shown in Figure 12.1. Note the difference between the histograms across the camera break between the second and the third frames, while the histograms of the first and second frames are almost identical. Figure 12.4 illustrates the application of histogram comparison to a documentary video. The graph displays the sequence of SD_i values defined by Equation 12.4 between every two consecutive frames over an excerpt from this source. The graph exhibits two high pulses that correspond to two camera breaks. If an appropriate threshold is set, the breaks can be detected easily.

Equation 12.4 can also be applied to histograms of individual color channels. A simple but effective approach is to use color histogram comparison [NT92]: Instead of gray levels, j in Equation 12.4 denotes a code value derived from

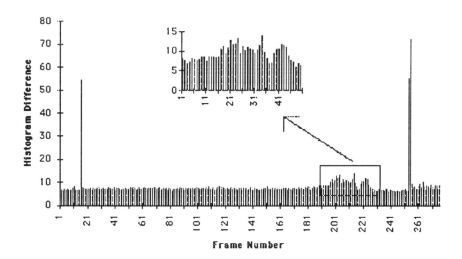

Figure 12.4 A sequence of frame-to-frame histogram differences obtained from a documentary video, where differences corresponding to both camera breaks and gradual transitions can be observed.

the three color intensities of a pixel. Of course, if 24 bits of color data were translated into a 24-bit code word, that would create histograms with 2^{24} bins, which is clearly unwieldy. Consequently, only the two or three most significant bits of each color component tend to be used to compose a color code. A six-bit code, concatenating the two high-order bits of each color intensity and providing 64 bins, has been shown to give sufficient accuracy [ZKS93]. This approach is also more efficient for color source material because it eliminates the need to first convert the color intensities into gray levels.

An alternative approach to linear histogram comparison is the following χ^2-test equation:

$$SD_i = \sum_{j=1}^{G} \frac{|H_i(j) - H_{i+1}(j)|^2}{H_{i+1}(j)} \qquad (12.5)$$

This metric was proposed in [NT92] for its ability to enhance differences between the frames being compared. However, experimental results reported in [ZKS93] showed that, while this equation enhances the difference between two frames across a camera break, it also increases the difference due to camera or object movements. Therefore, the overall performance is not necessarily better than that achieved by using Equation 12.4, while Equation 12.5 also requires more computation time.

Another approach to constructing a color histogram is to represent each pixel's coordinates in the HVC color space discussed in Chapter 11. One advantage of this approach is the invariance of Hue under different lighting conditions, as was observed in Chapter 11. As a matter of fact, [AHC93a] reports the use of a histogram based only on H and C coordinates. Rather than translating the color-coding approach of [NT92] to HC space, this alternative technique represents the histogram as a two-dimensional surface; and the difference between two histograms is also represented as a two-dimensional surface—a histogram of the differences for each corresponding pair of bins. In other words, if $h_i(H,C)$ is the histogram in HC space for frame i, then the difference between this frame and its successor, frame $i+1$, is also defined as a histogram, $\delta_i(H,C)$:

$$\delta_i(H,C) = h_i(H,C) - h_{i+1}(H,C) \qquad (12.6)$$

The overall difference is then calculated by the following volume formula:

$$\nu_i = \sum \{|\delta_i(H,C)| \times \Delta H \times \Delta C\} \qquad (12.7)$$

ΔH and ΔC are the resolutions of the Hue and Chroma values which provide the quantization of the HC histogram.

Threshold Selection

Selection of appropriate threshold values is a key issue in applying both of the above segmentation algorithms. Thresholds must be assigned that tolerate variations in individual frames while still ensuring a desired level of performance. A "tight" threshold makes it difficult for "impostors" to be falsely accepted by the system, but at the risk of falsely rejecting true transitions. Conversely, a "loose" threshold enables transitions to be accepted consistently, at the risk of falsely accepting "impostors." In order to achieve high accuracy in video partitioning, an appropriate threshold must be found.

The threshold t, used in pair-wise comparison for judging whether a pixel or region has changed across successive frames, can be easily determined experimentally; and it does not change significantly for different video sources. However, experiments have shown that the threshold T_b for determining a segment boundary using either pair-wise or histogram comparison varies from one video source to another [ZKS93]. For instance "camera breaks" in a cartoon film tend to exhibit much larger frame-to-frame differences than those in a "live" film. Obviously, the threshold to be selected must be based on the distribution of the frame-to-frame differences of the video sequence.

Several approaches to image segmentation are based on the thresholding of luminance or individual color intensities [Pra91]. Typically, selecting thresholds for such spatial segmentation is based on the histogram of the particular amplitudes being considered. The conventional approaches include the use of a single threshold, multiple thresholds, and variable thresholds. The accuracy of a single threshold selection depends upon whether the histogram is bimodal, while multiple threshold selection requires clear multiple peaks in the histogram. Variable threshold selection is based on local histograms of specific regions in an image. In spite of this variety of techniques, threshold selection is still a difficult problem in image processing and is most successful when the solution is application dependent. Nevertheless, setting an appropriate threshold for temporal segmentation of video sequences can draw upon an analogous feature, the histogram of frame-to-frame differences. Thus, it is necessary to know the distribution of the frame-to-frame differences across camera breaks.

A typical distribution of normalized frame-to-frame differences over an entire given video source exhibits a high and sharp peak corresponding to a large number of consecutive frames that have a very small difference between them. The distribution then tends to tail off for larger values, corresponding to the small number of consecutive frames between which a significant difference occurs. Because this histogram has only a single modal point, the approaches for threshold selection used in image segmentation tend not to be applicable; so an alternative approach [ZKS93] is necessary.

If there is no camera shot change or camera movement in a video sequence, the frame-to-frame difference value can only be due to three sources of noise: noise from digitizing the original analog video signal, noise introduced by video production equipment, and noise resulting from the physical fact that few objects are perfectly still. All three sources of noise can be assumed to be Gaussian. Thus, the distribution of frame-to-frame differences can be decomposed into a sum of two parts: the Gaussian noises and the differences introduced by camera breaks, gradual transitions, and camera movements. Obviously, differences due to noise have nothing to do with transitions. Statistically, the second part usually accounts for less than 15% of the total number of frames in a typical video [ZKS93].

Let σ be the standard deviation and μ the mean of the frame-to-frame differences. If the only departure from μ is due to Gaussian noise, then the probability integral

$$P(x) = \int_0^x \frac{1}{\sqrt{2\pi}\sigma} e^{-\frac{(x-\mu)^2}{2\sigma^2}} dx \qquad (12.8)$$

(taken from 0 since all differences are given as absolute values) will account for most of the frames within a few standard deviations of the mean value. In other words the frame-to-frame differences from the non-transition frames will fall in the range of 0 to $\mu + \alpha\sigma$ for a small constant value α. For instance $\alpha = 3$ in Equation 12.8 will account for 99.9% of all difference values. Therefore, the threshold T_b can be selected as

$$T_b = \mu + \alpha\sigma \qquad (12.9)$$

That is, difference values that fall out of the range from 0 to $\mu + \alpha\sigma$ can be considered indicators of segment boundaries. Experimental evidence [ZKS93] has shown that a good value for α tends to be between five and six. Under a Gaussian distribution the probability that a non-transition frame will fall out of this range is practically zero.

Motion Continuity

Another important property of any video source is the motion it captures. In psychology the term *flow* (or "optical flow," "flow perspective," or "streaming perspective") is used to designate the way in which different areas of the field of view detect motion in different directions [Gib86]. Optical flow may be represented quantitatively by assigning a field of motion vectors to the pixels of an image [HS81]. Techniques for computing motion vectors are discussed in Chapter 7.

Two approaches to using motion vectors for detecting camera breaks are proposed in [A+92]. Both assume that the motion vectors have been computed by *block matching*, which assigns to each block in frame i, b_i, a vector which displaces (i.e. translates) the entire block, D_{b_i}. The first approach involves correlating the displacement of b_i with the *actual* region in frame $i + 1$ occupying the displaced region. [A+92] proposes computing a product-moment coefficient of correlation with respect to the first moment (mean) in intensity values. However, a simpler approach would be to compute a normalized difference value similar to that used in pair-wise comparison. In either case a correlation value can be computed for each block, and these values can be averaged over all the blocks in the frame. If this average indicates a suitably low level of correlation, that lack of correlation may be interpreted as evidence of a camera break.

The second approach proposed in [A+92] involves computing a measure of *motion smoothness*. This is defined to be the "ratio of velocity to motion." The numerator is actually based on speed, rather than velocity. For each block, b_i, $w_{1_i}(b)$ is defined as follows:

$$w_{1_i}(b) = \left\{ \begin{array}{ll} 1 & \text{if } |D_{b_i}| > t_s \\ 0 & \text{otherwise} \end{array} \right. \tag{12.10}$$

Similarly, $w_{2_i}(b)$ is defined as follows:

$$w_{2_i}(b) = \left\{ \begin{array}{ll} 1 & \text{if } |D_{b_{i+1}} - D_{b_i}| > t_m \\ 0 & \text{otherwise} \end{array} \right. \tag{12.11}$$

t_s and t_m are both threshold values close to zero. Thus, $\sum_b w_{1_i}(b)$ counts the number of "significant" motion vectors in frame i; and $\sum_b w_{2_i}(b)$ counts the number of motion vectors in frame i which differ "significantly" from their corresponding vectors in frame $i + 1$. The smoothness of frame i, W_i, is then defined as the ratio of these sums:

$$W_i = \frac{\sum_b w_{1_i}(b)}{\sum_b w_{2_i}(b)} \tag{12.12}$$

This ratio will approach zero as more vectors change across successive frames, so a ratio which is less than a given threshold value may be interpreted as evidence of a camera break.

12.2.3   Gradual Transitions

The Twin-Comparison Approach

As one can observe from Figure 12.4, the graph of the frame-to-frame histogram differences for a sequence exhibits two high pulses that correspond to two camera breaks. It is easy to select a suitable cutoff threshold value (such as 50) for detecting these camera breaks. However, the inset of this graph displays another sequence of pulses the values of which are higher than those of their neighbors but are significantly lower than the cutoff threshold. This inset displays the difference values for the dissolve sequence shown in Figure 12.2 and illustrates why a simple application of this difference metric is inadequate.

The simplest approach to this problem would be to lower the threshold. Unfortunately, a lower threshold cannot be effectively employed, because the difference values that occur during the gradual transition implemented by a special effect, such as a dissolve, may be smaller than those that occur between the frames within a camera shot. For example, object motion, camera panning, and zooming also entail changes in the computed difference value. If the cutoff threshold is too low, such changes may easily be registered as "false positives." The problem is that a single threshold value is being made to account for all segment boundaries, regardless of context. This appears to be asking too much of a single number, so a new approach has to be developed.

In Figure 12.2 it is obvious that the first and the last frame are different, even if all consecutive frames are very similar in content. In other words the difference metric applied in Section 12.2.2 with the threshold derived from Figure 12.4 would still be effective were it to be applied to the first and the last frame directly. Thus, the problem becomes one of detecting these first and last frames. If they can be determined, then each of them may be interpreted as a segment boundary; and the period of gradual transition can be isolated as a segment unto itself. The inset of Figure 12.4 illustrates that the difference values between most of the frames during the dissolve are higher, although only slightly, than those in the preceding and following segments. What is required is a threshold value that will detect a dissolve *sequence* and distinguish it from an ordinary camera shot. A similar approach can be applied to transitions implemented by other types of special effects. This *twin-comparison* approach will be presented in the context of an example of dissolve detection using Equation 12.4 as the difference metric.

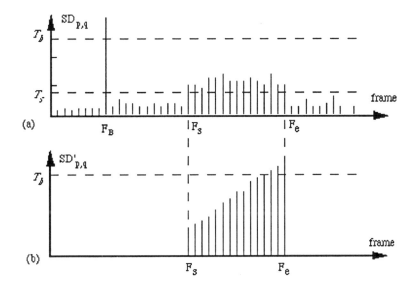

Figure 12.5 Illustration of twin-comparison: $SD_{p,q}$, the difference between consecutive frames defined by the difference metric; $SD'_{p,q}$, the accumulated difference between the current frame and the potential starting frame of a transition; T_S, the threshold used to detect the starting frame (F_S) of a transition; T_b, the threshold used to detect the ending frame (F_e) of a transition. T_b is also used to detect camera breaks and F_B is such a camera break. $SD'_{p,q}$ is only calculated when $SD_{p,q} > T_S$.

Twin-comparison requires the use of two cutoff thresholds: T_b is used for camera break detection in the same manner as was described in Section 12.2.2. In addition a second, lower, threshold T_S is used for special effect detection. The detection process begins by comparing consecutive frames using a difference metric such as Equation 12.4. Whenever the difference value exceeds threshold T_b, a camera break is declared, e.g., F_B in Figure 12.5a. However, twin-comparison also detects differences that are smaller than T_b but larger than T_S. Any frame that exhibits such a difference value is marked as the potential start (F_S) of a gradual transition. Such a frame is labeled in Figure 12.5a. This frame is then compared to subsequent frames, as shown in Figure 12.5b. This is called an *accumulated comparison* since, during a gradual transition, this difference value will normally increase. The end frame (F_e) of the transition is detected when the difference between consecutive frames decreases to less than T_S, while the accumulated comparison has increased to a value larger than T_b.

Algorithms used	N_d	N_m	N_f	N_P	N_Z
Gray-level comparison	101	8	9	2	1
χ^2 gray-level comparison	93	16	9	2	3
Color code comparison	95	14	13	1	2

Table 12.1   Detecting gradual transitions with three types of twin-comparison algorithms applied to a documentary video: N_d, the total number of transitions correctly detected; N_m, the number of transitions missed; N_f, the number of transitions misdetected; N_P, the number of "transitions" actually due to camera panning; N_Z, the number of "transitions" actually due to camera zooming.

Note that the accumulated comparison is only computed when the difference between consecutive frames exceeds T_S. If the consecutive difference drops below T_S before the accumulated comparison value exceeds T_b, then the potential start point is dropped and the search continues for other gradual transitions. The key idea of twin-comparison is that two distinct threshold conditions be satisfied at the same time. Furthermore, the algorithm is designed in such a way that gradual transitions are detected *in addition* to ordinary camera breaks.

A problem with twin-comparison is that there are some gradual transitions during which the consecutive difference value *does* fall below T_S. This problem is solved by permitting the user to set a tolerance value that allows a number (such as two or three) of consecutive frames with low difference values before rejecting the transition candidate. This approach has proved to be effective when tested on real video examples [ZKS93].

Some experimental results which summarize the efficacy of twin-comparison are summarized in Table 12.1 [ZKS93]. These figures are based on applying three difference metrics to the documentary video from which the images in Figure 12.1 and Figure 12.2 were taken. The first two rows of Table 12.1 show the results of applying difference metrics 12.4 and 12.5, respectively, to gray-level histograms. The third row shows the result of applying difference metric 12.4 to histograms of the six-bit color code. The video for this particular example was sampled at fifteen frames per second, so every second frame was compared from the original source.

Table 12.1 shows that histogram comparison, based on either gray-level or color code, gives very high accuracy in detecting gradual transitions. It is equally effective in detecting camera breaks, but those numbers have not been included in Table 12.1. In fact, even though no effort was made to tune the thresholds to obtain these data, approximately 90% of the breaks and transitions are correctly detected. Color gives the most promising overall results: while it is less accurate than gray-level comparison in detecting gradual transitions, it is more accurate for simple camera breaks, as well as being the fastest of the three algorithms [ZKS93]. (The 6-bit color code requires only 64 histogram bins, instead of the 256 bins required for the 8-bit gray-level.)

The rightmost two columns in Table 12.1 account for transitions detected by twin-comparison which are actually due to camera operation, another source of gradual change in image content. Distinguishing these two classes of transitions requires an analysis of motion content. This analysis will be discussed in Section 12.3.1.

Mathematical Models of Gradual Transitions

Once the intervals of gradual transition have been *extracted*, it may still be necessary to *classify* them. Twin-comparison does not provide the information which, for example, distinguishes a fade from a dissolve. [HJW94] presents a mathematical model of how frame images are transformed over the course of a transition between camera shots. The foundation of this model is a representation of an image as a mapping from a two-dimensional coordinate space into a vector in RGB color space:

$$E(x, y) = (r, g, b) \tag{12.13}$$

Sequences of images (such as camera shots and the transitions between them) may then be modeled by adding a temporal index, t, which is actually quantized in terms of frame number:

$$E(x, y, t) = (r, g, b) \tag{12.14}$$

Let S_1 and S_2 be two camera shots, represented as such mappings. Then the mapping for the transition between them $E_{1,2}$ may be defined as follows:

$$E_{1,2}(x, y, t) = T_{c1}(S_1(T_{s1}(x, y, t))) \otimes T_{c2}(S_2(T_{s2}(x, y, t))) \tag{12.15}$$

The T_{si} are transformations in pixel space, the T_{ci} are transformations in color space, and $\otimes$ represents the operator which actually combines the transformed data from the two shots. If both the pixel space and color space vectors are represented as *homogeneous coordinates*, then the T_{si} and T_{ci} can be represented as *affine two-dimensional transforms* [HJW94]. $\otimes$, on the other hand, often can be represented as simple addition.

This decomposition is useful in that it provides the basis for a typology of different transitions between shots. Thus, if both the T_{si} and T_{ci} transformations are identity transforms, then the transition is a simple camera break. If the T_{si} transformations are identity transforms, then the transition is strictly a manipulation of intensity space, such as a fade or a dissolve. If the T_{ci} transformations are identity transforms, then the transition involves pixel displacement, such as the emulation of a turning or sliding page. Finally, applying both sets of transformations implements more sophisticated transitions, such as morphing [HJW94].

Detecting and classifying transitions thus reduces to the problem of identifying the application of T_{si} and T_{ci} transformations, characterizing those transformations more specifically, and identifying the $\otimes$ operator which combines the transformations of the two camera shots. Unfortunately, the solution of this problem in the most general case is no easy matter. Thus, when it comes to simply *detecting* the presence of gradual transitions, the actual performance of this mathematical analysis [HJW94] falls significantly short of the capabilities of twin-comparison [ZKS93].

Multi-pass Approach

Once threshold values have been established, the partitioning algorithms can be applied in "delayed real-time." Only one pass through the entire video package is required to determine all the camera breaks. However, the delay can be quite substantial given 30 frames per second of color source material. In addition such a single-pass approach has the disadvantage that it does not exploit any information other than the threshold values. Therefore, this approach depends heavily on the selection of those values.

A straightforward approach to the reduction of processing time is to lower the resolution of the comparison. This can be done in two ways—either spatially or temporally. In the spatial domain one may sacrifice resolution by examining only a subset of the total number of pixels in each frame. However, this is clearly risky since if the subset is too small, the loss of spatial detail may result in a failure to detect segment boundaries. Also, resampling the original image may even increase the processing time, thus defeating the original goal.

Alternatively, sacrificing temporal resolution by examining fewer frames is a better choice, since in motion video temporal information redundancy is much higher than spatial information redundancy. An example of this approach has already been given in the experimental data presented in Table 12.1, since only

15 frames per second of the 30-frame-per-second source video were examined. This "skip factor" of two may even be increased for other video sources. Indeed, if the skip factor is large enough, then some gradual transitions will actually be detected as camera breaks, since the difference between frames across the skip interval may be larger than the threshold T_b. A drawback of this approach is that the accuracy of locating the camera break decreases with the size of the skip. Also, if the skip factor is too large, the change during a camera movement may be so great that it leads to a false detection of a camera break [ZKS93].

These problems may be overcome with a multi-pass approach that improves processing speed and achieves the same order of accuracy. In the first pass resolution is sacrificed temporally to detect *potential* segment boundaries. In this process twin-comparison for gradual transitions is not applied. Instead, a lower value of T_b is used; and all frames across which there is a difference larger than T_b are detected as potential segment boundaries. Due to the lower threshold and large skip factor, both camera breaks and gradual transitions, as well as some artifacts due to camera movement, will be detected; but any number of false detections will also be admitted, as long as no *real* boundaries are missed. In the second pass all computation is restricted to the vicinity of these potential boundaries. Increased resolution is used to locate all boundaries (both camera breaks and gradual transitions) more accurately.

With the multi-pass approach different detection algorithms can be applied in different passes to increase confidence in the results. For instance a video source may be analyzed with either pair-wise or histogram comparison in the first, low resolution, pass with a large skip factor and a low value of T_b. Then, in the second pass, *both* comparison algorithms may be applied independently to the potential boundaries detected by the first pass. The results from the two algorithms can then be used to verify each other, and positive results from both algorithms will have sufficiently high confidence to be declared as segment boundaries. An example of this "consensus" approach will be discussed in the case study in Chapter 14.

12.3 REPRESENTATION AND CLASSIFICATION OF CAMERA SHOTS

Thus far we have only addressed the problems of representation and classification as they apply to transitions between camera shots. However, the content of a video program almost never resides in these transitions. Content is carried by the shots themselves, and the transitions are simply creative devices skillfully applied to direct and hold the attention of the viewer [BT93]. Identifying content is thus fundamentally a problem of *perception*; and, unfortunately, perception is ultimately an act of *interpretation* on the part of the perceiver, an act which is inescapably *subjective* [Hus70].

Whether or not a computer can ever possess the consciousness which enables such subjective interpretation is a question of artificial intelligence which is far beyond the scope of this book [Ede89]. In the interests of pragmatic engineering, such subjectivity should remain the responsibility of the user, leaving the computer to process image and video data objectively in manners which will make it easier for the user to bear that responsibility. This leads to a technique which might be called "content-free content analysis." This technique involves using relatively straightforward algorithms and methods to furnish the user with appropriate *cues* to the content of a video, usually in the form of appropriate displays.[2] Combined with appropriate operators for manipulation, these cues form the basis of a tool set which the user may apply in the task of content analysis.

The primary function of any tool set is to relieve the user of the burden of the routine. Given a tool set which makes it easier for a user to get through a tedious task, such as logging the contents of a long video source, it may be possible to discover that there are now higher-level tasks being executed by the user which are *also* routine (although that routine nature could not have been discovered until tools were present to eliminate the routine at a lower level). The identification of routine at a higher-level may then lead to another generation of tools and user interfaces. In the course of time, such tools may strip away from the user the need to worry about anything other than the most subjective issues of interpretation.[3]

---

[2] These cues may play a role similar to those *affordances* provided by the environment which lie at the heart of James Gibson's "ecological" approach to perception [Gib86].

[3] Such an approach to tool development may be expressed by the aphorism: "Laziness is the mother of invention."

The first generation of such tools will still concentrate on a very low (and extremely objective) level of content analysis. Tools are currently available which provide useful visual cues for a variety of content-related properties. The capabilities of these tools include the construction of static images which serve to "abstract" the contents of a camera shot, the representation of camera operation, and the analysis of motion within a shot. Specific techniques which enable such tools to be implemented will now be discussed.

12.3.1   Camera Operation Detection

Given the ability to detect gradual transitions such as those that implement special effects, one must distinguish changes associated with those transitions from changes that are introduced by camera panning or zooming. Changes due to camera movements tend to induce successive difference values of the same order as those of gradual transitions, so that the problem cannot be resolved by introducing yet another cutoff threshold value. Instead, it is necessary to detect patterns of image motion that are induced by camera operation.

Motion Vector Analysis

Patterns of image motion associated with camera movement are most evident in fields of motion vectors, such as those discussed in Section 12.2.2. The optical flow fields resulting from panning and zooming are illustrated in Figure 12.6. In contrast, transitions implemented by special effects, such as dissolve, fade-in and fade-out, will not induce such motion fields. Therefore, if such motion vector fields can be detected and analyzed, changes introduced by camera movements can be distinguished from those due to special-effect transitions.

As illustrated in Figure 12.6, during a camera pan these vectors will predominantly have the same direction. (Clearly, if there is also object movement in the scene, not all vectors need share this property.) Thus, the distribution of motion vectors in an entire frame resulting from a camera panning should exhibit a strong modal value that corresponds to the movement of the camera. In other words most of the motion vectors will be parallel to the modal vector. This may be expressed in the following inequality:

$$\sum_{b=1}^{N} |\theta_b - \theta_m| \leq \Theta_p \tag{12.16}$$

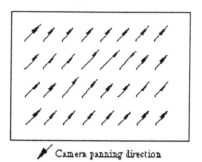

Camera panning direction

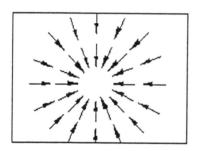 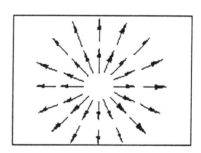

Figure 12.6 Motion vector patterns resulting from camera panning and zooming: the upper image illustrates panning and tilting; the lower images depict zoom out (left) and zoom in (right).

θ_b is the direction of the motion vector for block b, D_b, as described in Section 12.2.2; and N is the total number of blocks into which the frame is partitioned. θ_m is the modal value of the directions of the entire set of motion vectors. The difference $|\theta_b - \theta_m|$ is zero when the vector for block b has exactly the modal direction. The sum thus counts the total variation in direction of all motion vectors from the direction of the modal vector, and a camera pan is declared if this variation is smaller than Θ_p. Ideally, Θ_p should be zero; and a non-zero threshold is necessary to accommodate errors such as those that may be introduced by object motion.

In the case of zooming, the field of motion vectors has a minimum value at the focus center: focus of expansion (FOE) in the case of zoom out, or focus of contraction (FOC) in the case of zoom in. Indeed, if the focus center is located in the center of the frame and there is no object movement, then the mean of all the motion vectors will be the zero vector. Unfortunately, this is not always the case; and locating a focus center is not always an easy matter.

One way to circumvent this problem is to determine zooming on the basis of "peripheral vision." This approach only requires the assumption that the focus center lie within the boundaries of a frame. It requires comparing the vertical components of the motion vectors for the top and bottom rows of a frame, since during a zoom these vertical components will have opposite signs. Mathematically, this means that in every column the magnitude of the difference between these vertical components will always exceed the magnitude of both components:

$$|v_k^{top} - v_k^{bottom}| \geq max(|v_k^{top}|, |v_k^{bottom}|) \tag{12.17}$$

One may again modify this "always" condition with a tolerance value; but this value can generally be low, since there tends to be little object motion at the periphery of a frame. The horizontal components of the motion vectors for the left-most and right-most columns can then be analyzed the same way. That is, the vectors of two blocks located at the left-most and right-most columns but at the same row will satisfy the following condition:

$$|u_k^{left} - u_k^{right}| \geq max(|u_k^{left}|, |u_k^{right}|) \tag{12.18}$$

A zoom may then be said to occur when both conditions 12.17 and 12.18 are satisfied for the majority of the motion vectors.

The Video X-Ray

Another way in which to detect camera operation is by the examination of what are known as *spatiotemporal* images [AB85]. Figure 12.7 illustrates how such an image may be constructed. Just as an image may be represented as an *array* of pixels, a video may be represented as a *rectangular solid*. When represented this way, this solid has faces, one of whose dimensions is the time domain. Thus, the upper face of Figure 12.7 illustrates the changes in the top row of pixels of each frame over the course of time.

Figure 12.7   Constructing a spatiotemporal image.

Such images constitute what has been called a *video X-ray* [T+93]. In Figure 12.7 the texture of lines parallel to the time axis is a sure sign of a camera being held steady. (Note that, once again, by examining the *periphery* of the image, the likelihood of interference from object motion tends to be minimized.) On the other hand a texture of lines which are parallel to each other, but not to the time axis, will indicate panning, tilting, or a combination of the two, depending on whether these lines are observed in the top face, the side face, or both. Finally, a texture of lines which systematically converge or diverge, and are observed to do so on both the top and side faces, provide evidence of a zoom out or in, respectively. This approach constitutes a potential advantage over motion vector analysis, particularly if the computation of the motion vectors is too time-consuming.

The Hough Transform

Another way to analyze the motion vectors of a video frame is to map them all into a polar coordinate space using the Hough transform [Pra91]. As illustrated in Figure 12.8, any line in the x-y plane can be described parametrically as

$$\rho = x \cos \theta + y \sin \theta \tag{12.19}$$

ρ is the normal distance from the origin to the line, and θ is the angle of that particular normal vector with respect to the x axis. Since every motion vector is a segment of a line, the Hough transform maps each vector into a point in polar coordinate space (sometimes called Hough space).

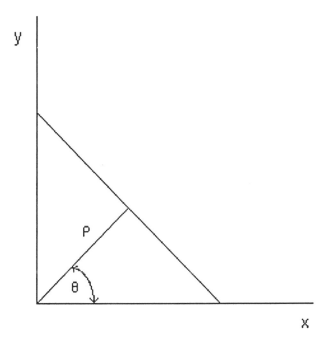

Figure 12.8 Parametric representation of a line in polar coordinates.

Motion vectors whose directions exhibit a strong modal value will appear in Hough space as a set of points sharing approximately the same θ value. Thus, any combination of panning and tilting will be manifest in Hough space as a straight line perpendicular to the θ axis [A+92]. On the other hand a family of lines, all of which intersect at a common point, will plot in Hough space as points along a sinusoidal curve [Pra91]. The motion vectors associated with a camera zoom will be segments of lines with this property, so such a sinusoidal pattern in Hough space will be indicative of such camera operation [A+92].

From a mathematical point of view, it is unclear that representation in Hough space offers much advantage over statistical analysis. Indeed, given the properties of object motion, both approaches tend to benefit from concentrating on peripheral motion vectors. However, from the point of view of user analysis, Hough space offers a more accommodating presentation of motion vector data than does a vector field. Points plotted in Hough space are more amenable to the detection of both patterns and outliers [A+92]. Thus, to the extent that effective classification will require constructive interaction between a user and a set of software tools, the Hough space display definitely can serve as an advantageous tool.

12.3.2   Salient Video Stills

Information from camera operation may also be used to construct a single, composite, static image. Thus, for example, if the camera slowly pans across a scene, a panorama of the entire scene may be constructed by combining portions of the frames associated with different stages of the pan. Similarly, information from a zoom can be used to enable the "broader view" to provide pixel information "beyond the edges" of the "narrower view." Composite images which are constructed in this manner are known as *saflient video stills* [TB93].

Of course the basic theory behind this approach works only if there is no object motion while there is camera motion. On the other hand object motion provides a means of revealing additional information about the background, since different portions of the background will be obscured as the object crosses it. Consequently, the salient still may be constructed in such a way that the object appears in only one place; and the background is appropriately "fleshed out." Alternatively, if the entire background can be reconstructed, the still representation may be just as useful if the object is eliminated entirely.

12.3.3   Key Frames

Perhaps the easiest way to represent a camera shot is to "abstract" it as a set of images, extracted from the shot as representative frames, usually called *key frames* [O'C91]. Certainly, as a *manual* operation, key frame extraction is far simpler than constructing salient stills or video X-rays. However, even if the camera shots have already been identified, selecting key frames can still be a tedious process which would benefit from automation.

Fortunately, the process of key frame extraction *can* be automated quite robustly [ZLS95]. Furthermore, the process can be parameterized in such a way that the number of key frames extracted for each shot depends on features of the shot content, variations in those features, and the camera operations involved. The technique is based on the use of a selected difference metric, similar to those discussed in Section 12.2.2. However, in this case all computation takes place within a single camera shot and differences between consecutive frames are not computed. Instead, the first frame is proposed as a key frame; and consecutive frames are compared against that candidate. A two-threshold technique, similar to the twin-comparison method discussed in Section 12.2.3, is applied to identify a frame significantly different from the candidate; and that frame is proposed as another candidate, against which successive frames are compared. Users can specify the density of the detected key frames by adjusting the two threshold values.

12.3.4 The VideoMAP

Graphs such as the one depicted in Figure 12.4 can also provide useful representations of video content. The *VideoMAP* is a display technique developed at the NTT (Nippon Telephone and Telegraph) Human Interface Laboratories which takes advantage of such representations [T⁺93]. This display provides spectrogram-like renditions of histograms for both gray-level and Hue. It also presents a plot of the average intensity level for each frame and a plot of successive frame differences as in Figure 12.4. Finally, the display provides the X-ray views discussed in Section 12.3.1. The VideoMAP thus offers an interesting presentation of multiple abstract representations of the information contained in a camera shot.

12.3.5 Visualizing Statistics

Another approach to representing the content of camera shots is in terms of statistical properties. The case study reviewed in Chapter 14 will discuss how shots with relatively static spatial structure, such as anchor desk shots in news broadcasts, may be "averaged" over their duration. Thus, a shot of an anchorperson should reveal minimal body movement, except for the lips. Consequently, a simple averaging of pixel values for each frame of the shot enables the synthesis of an "average image" for the entire shot. This image will be relatively clear except for areas which are blurred due to motion. If one is trying to recognize who a particular anchorperson is, that blurring is not likely to impact the identification task.

Any such averaging attempt depends heavily on how representative the average is of the data being averaged. Statistically, this is generally captured by the variance of the samples. Consequently, a display of an average image can be further enhanced to display a "variance temperature:" the smaller the variance value, the more representative is the average image. Naturally, higher variance will lead to higher blurring in the average image; but the "temperature" display provides an additional approach to interpreting that blurring.

12.4   INDEXING AND RETRIEVAL

Chapter 11 discussed the indexing and retrieval of static images. The major problem introduced by video source material is its considerable volume. Digitizations of individual videos are just too large to be feasibly stored in today's databases. Consequently, a database must be constructed with respect to one or more *abstractions* of a video's content. The most familiar abstraction is that of text—the sort of information which tends to be provided by a *log*. However, as databases become more multimedia, abstractions based on static images become more feasible. Other approaches to abstraction include information about camera operation or motion within individual camera shots. Each of these abstractions can play a role in how a description of a video may be suitably indexed for subsequent retrieval. Nevertheless, consulting a video resource is not always a matter of well-focused retrieval. Often browsing can be just as important, if not more so; so a database of video content should be able to support browsing as readily as it supports retrieval. These issues will now be discussed in greater depth.

12.4.1   Text-Based Abstractions

Stratification

The simplest way in which to describe the content of camera shots is with text. However, if it is assumed that multiple users will be examining this text and trying to extract information from it, then it is useful to impose stylistic conventions on how that text will be written. To some extent these conventions will reflect assumptions as to how video source material is perceived, since those assumptions will acknowledge, among other things, what does and does not get described.

What makes video different from static images is the *passage of time*. Thus, in describing a single image, attention tends to be focused on the *objects* in that image and how they are spatially related. However, the time dimension introduces the concept of an *event*; and events admit of different descriptions than objects. More specifically, events are associated with temporal intervals. Those intervals must be described in terms of *duration*, and any event is temporally related to other events with respect to *sequential ordering* and *simultaneity*. Thus, text which describes video content must also be capable of describing events in these terms.

Within a single camera shot there may be many events. Different people are doing different things; a car may drive by or an airplane fly overhead. There will also be *audio* events whose sounds may not always correspond to what the camera sees. Thus, a shot of a volleyball game at the beach may concentrate on the game; but there will probably also be the sound of the sea in the background, even if the camera is facing away from the water. Similarly, the video may record the sound of a jet plane flying by without the plane actually entering the field of view.

An event-based text description must thus function somewhat like a schedule, accounting for all events with respect to both what they are and when they occur. Such schedules are frequently easier to read and comprehend if the verbal description is supplemented with a graphic presentation. An example of such a presentation is given in Figure 12.9. This kind of display is sometimes called a " Stratagraph," since it serves to illustrate an approach to video annotation known as the "Stratification method" [AS92]. The Stratification method explicitly confronts the need to account for all events in a camera shot as independent entities, each of which requires its own description, rather than reducing the description of the shot to a summarizing "global" description.

The major problem with this kind of description is that its level of detail makes it extremely time-consuming. Consequently, the Stratagraph is more than a passive display. Instead, it is part of a suite of interactive tools which facilitate the task of annotation based on identifying, labeling, and describing component events within a camera shot [AS92]. The volume of descriptive text is still high; but these tools make it easier to sort out objects as image clips, audio clips, or free text descriptions. The objective of the tools is to provide automatically those elements of the annotation which can be extracted by machine analysis,

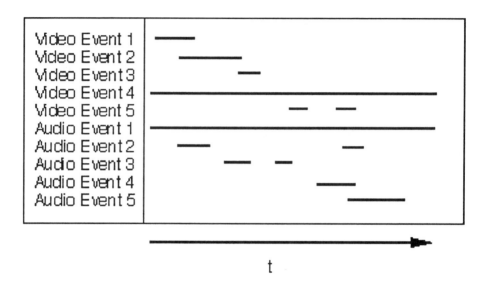

Figure 12.9   Representing the events in a camera shot as a schedule.

allowing the user to concentrate on only the more subjective demands of the descriptive task.

Object-Oriented Frames

Another approach to systematic text description is to regard a *database* as a highly constrained presentation of text material. Consequently, information concerned with the content of video source material which is maintained in a database [SSJ93] would serve as a well-structured presentation of text descriptions. This returns us to the question of how *indexes* will be built which facilitate retrieval from such a database. Chapter 11 demonstrated that it is possible to base such an index entirely on low-level features of image content, but it is clear that retrieval tasks will also benefit from some representation of *semantic* properties of the content [PPS94]. With respect to the task of index construction, perhaps the most important semantic property is classification with respect to a collection of topical categories—the property which governs the construction of most book indexes. Such an index may be implemented as a frame-based system in order to represent such topical categories as a class hierarchy [SZW94]. This hierarchical structure would reflect the indented structure found in the indexes of many books.

In frame-based systems the most important impact of a representation based on a hierarchy of classes is the contribution of that hierarchy to *inheritance* . Objects which are lower in the hierarchy are interpreted as *specializations* of their ancestors. As such they share (inherit) the descriptive properties of those ancestors but extend those descriptions with additional detail which makes them unique as specializations. In frame-based systems such inheritance applies not only to propositional descriptions but also to procedures (sometimes called "demons") which may be *attached* to different objects in the class hierarchy and embody specific operational knowledge regarding how instances of those classes may be manipulated, created, or destroyed. Thus, inheritance has both *descriptive* and *operational* impact on the units of knowledge (records) which are stored or examined in a database [BF81]. As an example of how both aspects of inheritance contribute to the construction of an index, let us consider the problem of structuring a database which maintains an archive of stock footage— material explicitly stored for use in a variety of different contexts.

Structural Foundations A major reason one may wish to accumulate video resources in a database is that they may be *reused* in subsequent production work [Dav93]. (Newspapers often tend to save copies of all of their photographs for similar purposes.) Indeed, there is now a major commercial business concerned with the archiving and distribution of such stock footage which can be made available to a broad variety of clients, each of whom may be interested in the material for radically different production purposes.

Such archives clearly need indexing . Traditionally such collections are often organized only in the head of a single archivist, who can satisfy any request by remembering exactly where every item is and what it contains. Unfortunately, such specialized knowledge is rarely shared; when an individual like that archivist leaves, all the organizational information leaves with him. What is required is an indexing system which embodies this organization information but is, at the same time, flexible enough to accommodate users with a wide variety of needs, each of whom may look at a given source of material in a different way [PPS94].

The difficulty is that this source material is, by both its nature and the way in which it is used, fundamentally *unstructured*. There are no structural cues which will facilitate the definition of organizational categories. There are only individual camera shots, none of which need exhibit any meaningful relationships among each other and any of which may be so related in the mind of a producer seeking just the right material for a particular creative act.

Class Organization The problem, then, is that any attempt at indexing will involve the classification of objects which inherently resist classification. Ultimately, there is only *one* category: StockShot. If we are to define this category as a frame, then our only hope for structure will lie in our ability to use the *slots* [BF81] of that frame to describe structural attributes and relationships. Furthermore, the expressiveness of the slot structure should also facilitate the problem of searching a very large collection of instances of a single class.

Slot Structure Slots provide a structural representation of the information about content required to *annotate* individual shots [BF81]. Some of this annotation must account for the lowest level of bookkeeping. Thus, a **Source** slot is required to provide a label of the source tape containing a particular shot; and **In** and **Out** slots are necessary to give the time codes of the beginning and ending of the shot itself. Any remaining slots may then be employed to describe the actual content of the shot.

One way to represent content is with the images of representative frames, as was discussed in Section 12.3.3. However, if text is also to be employed, then the danger is that free text or keywords may be too general-purpose to support searching and browsing. Instead, descriptive information is likely to be more useful if it is distributed over more specific slots, as in the following slot structure:

- **Source**

- **In**

- **Out**

- **RepresentativeFrame**

- **SourceType**

- **ForegroundActor**

- **BackgroundScene**

- **Activity**

The **SourceType** serves to characterize the nature of the image source: Is it live footage or synthesized graphics? Are the graphics cartoon depictions, or are they more abstract animations? **ForegroundActor** accounts for one or more agents who constitute the "center of attention" in a particular shot. They may be human actors from live footage or cartoon characters. There may also be shots, such as panoramas, which do not include *any* such agents. Such background information is captured by the **BackgroundScene** slot, which may also be left empty if the agents in the foreground are depicted without any identifiable background. Finally, there is the **Activity** itself, which is depicted by the shot. Again, this slot may be empty, as would be the case with a shot which simply surveys a particular background scene.

These specific slots are based on an examination of several hours of stock shot tapes. They do not necessarily constitute a complete set of descriptors for such shots. Indeed, it is quite possible that specific applications of this material would require additional slots. One advantage of most frame representation systems is that it is relatively easy for a user to supply additional slots, although such a user would then also have to establish how those slots were filled. Nevertheless, the representation system has the advantage of accommodating different needs of different users.

Demons   The primary role played by demons is to enforce a requirement that all slots be *strongly typed*: A value cannot be added to any slot unless it satisfies some data typing requirement [SZW94]. For example this requirement insures that a value added to either the **In** or **Out** slot is a well-formed time code and that the **Out** time occurs *after* the **In** time. However, an object-oriented environment can deal with classes as if they were, themselves, data types [SZW94]. Consequently, it is possible to define class hierarchies which characterize how specific descriptive slots may be filled; so let us see how such hierarchies may be defined for the four "subjective description" slots outlined above.

We have already reviewed the options currently being considered for the **SourceType** slot. Fillers of this slot may be typed by a **Source** class which serves as the root of the class hierarchy illustrated in Figure 12.10. Similarly, the **ForegroundActor** slot must be filled with instances of the **Actor** class. This class supports a very rich hierarchy of subclasses which will evolve as more instances of stock footage are examined. Figure 12.11 illustrates a representative subset of this hierarchy. Because we are dealing with a general-purpose data typing, we do not aspire to a sub-hierarchy of **Animal** which would necessarily reflect a biological taxonomy. On the other hand an informal categorization which respects some of the more technical categories one might find in zoological references, such as *Walker's Mammals of the World* [NP83], is likely to

```
Source
     LiveFootage
     Graphics
            Cartoon
            Animation
```

Figure 12.10   The Source class hierarchy.

```
Actor
     Human
            Adult
                   Man
                   Woman
            Child
                   Boy
                   Girl
     Animal
            Alligator
            Bee
            Monkey
```

Figure 12.11   The Actor class hierarchy.

be appropriate as a way to capture information about similarity during query processing.

The Scene class which types fillers of the **BackgroundScene** slot also supports a very rich hierarchy which will grow as more examples are accumulated, probably at a rate comparable to the Actor hierarchy. Figure 12.12 again provides a representative subset.

Similarly, the Action class which types fillers of the **Activity** slot is likely to grow, although it is unclear how rich it will be as a hierarchy. One approach, illustrated in Figure 12.13, is simply to enumerate different types of activities. This may prove to be one of the hardest classes to characterize suitably, particularly if Narrative is used as a "catch-all" when no other descriptor is appropriate. (For example, a cartoon sequence of a bee dive-bombing a dog can be classified as having a Narrative filler for its **Activity** slot.) Nevertheless,

```
Scene
     Nature
     Commercial
          Outdoor
          Indoor
```

Figure 12.12 The Scene class hierarchy.

```
Action
     Business
     Exercise
     Narrative
     Recreation
```

Figure 12.13 The Action class hierarchy.

the philosophy behind this approach is that each descriptor should be associated with a framework of guidelines for the terms which may be used; and a class hierarchy provides a useful means of implementing those guidelines.

Impact on Query Processing The focus of query processing is thus on slot-based retrieval . Each query is a set of *probes* which may be combined by either conjunction or disjunction operators. Each probe, in turn, requires the identification of a specific slot (which may be chosen from a menu). Once the slot has been selected, the user is then presented with the class hierarchy of its fillers and can choose any level of generality or specificity in that hierarchy. This approach is very similar to query processing interfaces for keyword-based databases [See92]; but it takes advantage of having semantic hierarchies for the fillers of each slot. This provides far more control over how specific one wishes to be in formulating a query than may be allowed by a logical disjunction of keywords. Such control can be particularly powerful if that disjunction requires a very large number of terms, as might be the case with, for example, a search for a general class of animal, such as mammals.

Data Modeling

Many of the insights associated with the use of object-oriented frames to represent index structures can be found in current database technologies. This is because for over a decade database research has directed a synthesis of concerns with knowledge representation and abstract data types towards an area now known as *conceptual* (or *semantic*) *data modeling* [MS81]. While earlier databases were specified solely in terms of *schemata* [Dat77] which basically designated how bits were organized into records, the trend has been to focus less on the *syntactic* priorities of schemata and more on the *semantic content* which a database is intended to embody.

As databases become multimedia, questions of semantics must address the content of the multimedia objects being stored. Thus, questions of data modeling now overlap heavily with those of content-based video indexing and retrieval. Two relevant database projects will now be examined to highlight the significance of this overlap.

VIMSYS VIMSYS is the Visual Information Management SYStem developed as part of the InfoScope project at the University of Michigan, a project concerned with various aspects of the development of heterogeneous databases [GWJ91]. The VIMSYS data model was developed with the following *desiderata* in mind:

- An image should be accessed either completely or in partitions.

- Image features should be considered both as independent entities and in their relations to the entire image.

- The image features should be hierarchically organized, so complex features may be represented as compositions of simpler ones.

- There should be several alternative ways in which a semantic feature may be derived from image features.

- The data model should support spatial data and file structures which infer spatial parameters associated with images and image features.

- The image features of complicated image regions should be represented in terms of nested or recursive definitions.

The result was a data model which provided four different "planes" along which the user could view the contents of a database:

DO: This is the plane of the Domain Objects and the relations among them.

DE: This is the plane of Domain Events and the relations among them; the representation of events enables the modeling of full-motion video, rather than just static images.

IO: This is the plane of Image Objects and the relations among them.

IR: This is the plane of Image Representations and the relations among them.

The most important aspect of this model is its distinction between the "raw" image data and the objects and events which those images depict (the domain). To say that an image may be interpreted in several different ways is to say that there are several different domains whose objects and events are being depicted in a common image. All that is important for the sake of consistency is that a single domain be associated with each interpretation.

Similarly, there is a distinction between the ontology of perceived objects in an image and that of the representation of those objects. The relationship between these two ontologies is what Gerald Edelman has called " perceptual categorization," the mechanism by which the brain takes the signals of visual sensation and establishes what objects there are in the world and how they are delimited [Ede87]. This is as much a question of interpretation as is the interpretation of image objects as domain objects. If the preceding paragraph addressed the problem of interpreting "raw" image objects, then it is also necessary to address the problem of interpreting "raw" stimuli as image objects in the first place! The significance of the VIMSYS model, then, lies in the number of degrees of freedom it provides to allow for maximum flexibility and variety in the interpretation of image and video data.

The Berkeley Distributed VOD System The Berkeley Distributed VOD (Video-On-Demand) System has addressed the problem of data modeling in terms of the specific problem of how multiple video servers can provide a wide variety of video source material to an equally wide variety of users [RBE94]. The modeling task was based on first interviewing a variety of users to identify the sorts of queries one might wish to pose in the course of searching for a video to "demand." These interviews revealed that there are three basic types of indexes which may be required to support those queries: bibliographic , structural , and content . Bibliographic indexes resemble the sorts of indexes we normally associate with library catalogs. Each entry would include a title, an abstract, keyword classifications of subject and genre, and the sort of information displayed in the opening and closing credits. Structural indexes are based on the sort of structural parsing described in Section 12.2. This would include identifying units of both temporal and spatial structure, along with explicit representation of any models guiding the structural analysis. The sort of content indexes envisaged by [RBE94] are based on the *objects of content* in a video source, including the appearance and location of persons and physical objects which serve as scenery or properties. This type of index should accommodate objects which are both visual and auditory; and it should incorporate different levels of description, such as a dominant color in a scene or a sound effect designated in terms of the physical object which produced it. However, there is a second type of content index which is closer to the sorts of indexes we normally associate with books. These are based on topical categories which reflect the subject matter of the source and should relate to the keyword terms included in related bibliographic indexes. An approach to integrating these three types of indexes into a common model is discussed in [Z+95].

12.4.2   Using Static Images

Several of the approaches discussed in Section 12.4.1 allowed for the possibility that a fundamentally text-based database could be supplemented by other media: images, video, and/or audio. Another approach is that the *text* content play the supplementary role and the database be structured about a suitable index of data in another medium. Chapter 11 addressed a variety of ways to take this approach for the medium of static images. It is important to note, however, that a database of static images can play other roles than that of a photographic archive. If the images in the database are key frames from a video, extracted by a process such as the one discussed in Section 12.3.3, then the database can provide a suitable abstraction of video content which may then be employed for most tasks concerned with retrieving the original video data. This means that the database need not address the problem of storing or managing the extremely large collections of data which constitute digitized video. Put another way, the processing of queries for a system such as the Berkeley Distributed VOD System would not require the resources of any of the video servers but could, instead, be managed by a database with significantly diminished storage requirements.

12.4.3   Abstractions Based on Camera Operation

Another approach is to use representations based on the detection of camera operation, such as those discussed in Section 12.3.1, as abstractions of video content. This abstraction is not necessarily extremely useful when applied on its own. However, as was observed in Section 12.3.1, camera operations frequently provide cues as to how the attention of the viewer should be directed. If the camera is zooming in, objects at the focus of the zoom are most likely being regarded by the director as important. When the camera zooms out, the focus is still important; but now the context within which that focal point is situated is also important. Thus, in both cases, the images at the beginning and ending of a zoom may be regarded as constituting an appropriate abstraction of the entire shot.

The situation is similar with panning and other realizations of global motion. Such movement takes the eye on a figurative "journey" from a "here" to a "there." In general, then, the entire journey can be abstracted into its beginning and end. Once again, images from the beginning and ending of the camera operation should serve as an abstraction of the entire transition sequence.

12.4.4 Motion-Based Abstractions

Section 12.4.1 discussed how VIMSYS includes domain events as part of its data model. However, this data model is primarily text-based. The SunSet Multimedia Information System wishes to deal with events, principally in the form of object trajectories, as a fundamental basis for abstraction, just as static images were discussed as a basis for abstraction in Section 12.4.2 [DG94]. In the spirit of VIMSYS, the approach is a hierarchical one with MPEG motion vectors at the bottom level. Vectors of successive frames are analyzed with the goal of identifying trajectory patterns which may then be interpreted as trajectories of objects in the image. However, the association of trajectories with objects requires further analysis based on knowledge of properties of the object. If the object is known to be rigid, it will have a single trajectory. On the other hand, if it is a linkage of rigid components, it will be associated with multiple trajectories, relations among which will be constrained according to the nature of the linkage. Once trajectories and objects have been identified, then specific properties of both trajectories and objects may be expressed in the formulation of queries.

Another motion-based abstraction of a camera shot is to view it as if it were an animation. Animations are generally constructed by decomposition into sets of transparent layers, often called *cells* (or "cels"). Background cells can be reused through long "camera shots," possibly being displaced to account for "panning." Foreground cells, on the other hand, change from frame to frame, representing the principal action. This is an alternative way to represent the association of movement patterns with component objects and thus provides another way to abstract the description of content. This technique has also been applied as a "semantic" approach to video compression and is described in [AW93].

12.4.5   Browsing Tools

While indexing and retrieval tend to be regarded as the major problems in developing databases and information systems which support video and other multimedia resources, *browsing* is another significant task which also merits quality computer-based support. By " browsing" we mean an informal perusal of content which may lack any specific goal or focus—Heidegger's "just asking around" [Hei77]. Such a perusal must respect the medium being browsed; in the case of video this means being able to see either action sequences or static images which are representative of the video's content. This task serves not only the general objective of "getting to know" a video source but also more focused query processing: The results of a database query, whether formulated in terms of text or visual features, will usually be a list of candidate video sequences which match the query criteria; and browsing is the best approach to examining that list. Video browsing will also be a major asset for video-on-demand systems, providing users with knowledge of what they are demanding before committing network resources to that request.

Unfortunately, the only major technological precedent for video browsing is the VCR (even available in "soft" form for computer viewing), with its support for sequential fast forward and reverse play. Browsing a video this way is a matter of skipping frames: the faster the play speed, the larger the skip factor. Because this skip factor is uniform, regardless of video content, it may be too coarse for some video sequences and too fine for others. Indeed, there is always the danger that some skipped frames may contain the content of greatest interest. Finally, there is the problem that high-speed viewing requires *increased* attention,[4] because of the higher data rate, while we prefer to associate browsing with a certain *relaxation* of attention. Consequently, any approach to browsing which is based on variations on such high-speed viewing, such as [A+94], may be more of a hindrance than a benefit.

---

[4] Think of what happens when you want to skip through the commercials and get back to the program you recorded!

In the spirit of this Chapter, the shortcomings of the VCR, either as a hardware device or as a metaphor for software, may be overcome by a more *content-based* approach to video browsing. In order to achieve this goal, effective browsing tools must facilitate movement between two levels of temporal granularity—overview and detail. In addition, they should support two different approaches to accessing video source data: *sequential* access and *random* access. What will make these tools content-based is their exploitation of various sources of knowledge discussed earlier in this Chapter, including segment boundaries, transition sequences, camera operations, and key frames of each segment.

Sequential access browsing usually entails a VCR-like interface like the one illustrated in Figure 12.14. Overview granularity is achieved by playing only the extracted key frames at a constant rate which is set by the user through a shuttle control. In this mode single-stepping advances through the key frames one at a time. Detailed granularity is provided by normal viewing, with frame-by-frame single stepping; and this approach is further enhanced to "freeze" the display each time a key frame is reached. A scroll bar is also provided to enable cuing to any location in the video; and the user can also enter frame numbers directly into the **From:** and **To:** windows in the upper right-hand corner. Finally, viewing at both levels of granularity is supported in reverse, as well as forward.

Random access may be provided by a *hierarchical browser* in which a video sequence is spread in space and represented by frame icons which function rather like a light table of slides. In other words the display space is traded for time to provide a rapid overview of the content of a long video. As shown in Figure 12.15, at the top of the hierarchy, a whole video is represented by five key frames, each corresponding to a segment consisting of an equal number of consecutive camera shots. Any one of these segments may then be subdivided to create the next level of the hierarchy. Unlike the Hierarchical Video Magnifier, which will be discussed in Section 12.5.2, this browser exploits structural knowledge of the video content, such as the location of shot boundaries. Descending through the hierarchy focuses attention on smaller groups of shots, single shots, the representative frames of a specific shot, and, only at the lowest levels, an unstructured sequence of frames represented by a key frame. We can then move to more detailed granularity by viewing sequentially any particular segment of video selected from this browser at any level of the hierarchy.

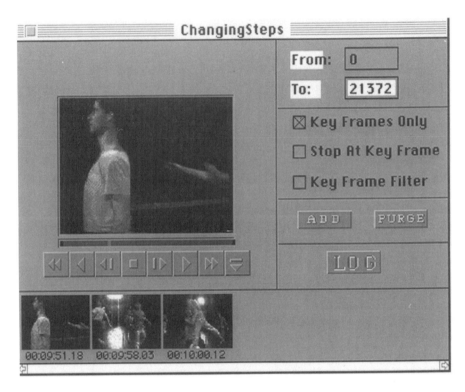

Figure 12.14   Content-based video player.

12.5   INTERACTIVE TOOLS

12.5.1   Micon Construction

An icon is a visual representation of some unit of information. If that information is time-related, then representing it as an icon requires being able to account for the temporal dimension. The *video icon* (also called a " micon" or "movie icon") is such an iconic representation of a video source [A+92]. This icon is constructed according to a very simple principle which was introduced in Section 12.3.1: If a single frame of video is represented by a two-dimensional *array* of pixels, then the video itself may be represented by a three-dimensional *volume*, where the third dimension represents the passage of time. Thus, the simplest way in which to construct a micon is simply to stack successive frames of a video, one behind the other, enabling the user to see the frame on the

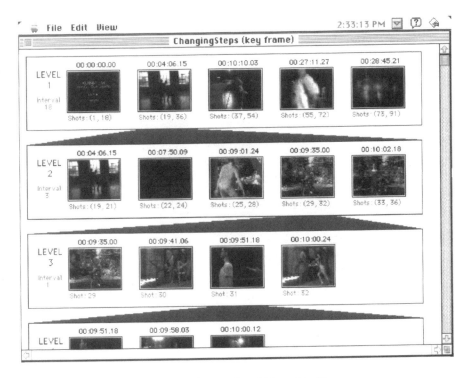

Figure 12.15 Content-based hierarchical browser.

front and the pixels along the upper and side faces. One example of such a construction was illustrated in Figure 12.7; another is given in Figure 12.16.

It is important to observe the extent to which the pixel traces recorded on the top and side faces of a micon can bear information as useful as frame images. The video source material for Figure 12.16 is a brief clip of the ascent of the rocket which launched the Apollo 11 mission to the moon. As the rocket rises, it cuts across the top row of pixels in each frame. The sequence of pixels in this row in successive frames thus contains a "trace" of the horizontal layers of the rocket which cross in each frame. Thus, a more complete image of the body of the rocket is reconstructed across the top face of the micon.

If a shot includes camera action, such as panning and zooming, then the spatial coordinates of a micon will no longer correspond to those of the physical space being recorded. However, a Hermart transform [A+92] may be used to construct a micon which is more like "extruded plastic" than a rectangular solid. As the camera zooms in and out, the size of the frame images shrink and expand,

Figure 12.16 A micon of the takeoff of Apollo 11.

respectively; and if the camera pans to the right, then the entire frame will be displaced to the right a corresponding distance. The resulting micon is then accurately embedded into its two spatial coordinates.

12.5.2 Interacting with Video Data

Iconic representations are only valuable to the extent that they permit useful interactions, and this is also true of video icons. We shall now consider how the video icon can be turned into an interactive tool. Two other interactive tools will also be discussed. One is a variation on the video icon, known as the video streamer; and the other uses static images to construct a hierarchical view of an entire video source.

Interactive Video Icons

Figure 12.17 illustrates an environment for the interactive examination and manipulation of micons, incorporating the image from Figure 12.16. The triangular button at the bottom of the column of operators on the left can be used to "play" the micon. It functions like the "play" button on a VCR; and the image of the "playing" video is displayed on the front face of the micon. The button above this "play" button is a "browse" button. When a micon is browsed, any point along a "depth face" can be used to cue the frame corresponding to that point in time for display on the front face. Thus, an icon need not be static but can represent its content through moving images [Ton91].

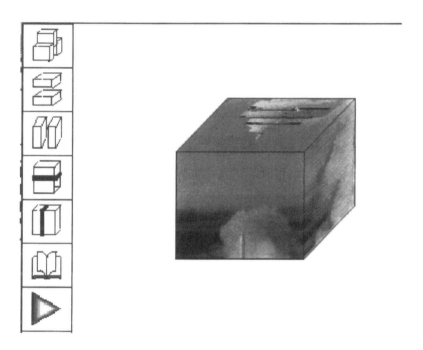

Figure 12.17 An environment for examining and manipulating micons.

The micon may be further examined by using the other operators indicated by the icons on the left side of the display in Figure 12.17. The simplest operations involve taking horizontal and vertical slices which allow for examination of pixels in the "interior" of the micon volume. Figure 12.18 shows an example of a vertical slice taken through an excerpt from the video *Changing Steps*, a "dance for television" conceived by choreographer Merce Cunningham and video designer Elliot Caplan. Just as the movement of the rocket left a trace of its own image on the top face of the micon in Figure 12.16, the sideways movement of one of the dancers translates into a distorted image of her body along the interior vertical face exposed by this slicing operation. The top operator is available to specify the beginning and ending frames of a sub-icon and extract that sub-icon by "lifting" it from the micon body. This is illustrated in Figure 12.19.

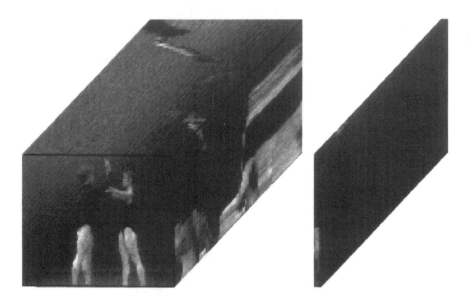

Figure 12.18   A vertical slice through a micon.

A Video Streamer

In addition to viewing a stored video clip as a volume of pixels, it is also possible to view a real-time display of video in the same manner. This type of display is known as a *video streamer* [Ell93], and it is similar in appearance to a micon. The only significant difference is the representation of time. In a micon the earliest event is at the front of the volume and increasing time proceeds "into the screen." In the video streamer the front face is always the most recent frame; so looking "into the screen" provides a "view of the past." Figure 12.20 is an example of a video streamer display.

The primary advantage of the video streamer is that it presents frames in the context of their recent past. This may allow the user to trace movements more

Figure 12.19   Extracting a sub-icon.

easily, particularly those which involve transitions across the edges of the frame. It also provides visual cues concerning transitions between camera shots. The display in Figure 12.20 includes a trace of a difference metric (in this case a histogram difference) on which the camera break threshold has been explicitly displayed.

The Hierarchical Video Magnifier

Sometimes it is more important to be able to browse a full-length video in its entirety than it is to be able to examine individual camera shots in detail. The Hierarchical Video Magnifier [MCW92] provides an approach to such browsing, and an example is illustrated in Figure 12.21. This is an overview of the entire *Changing Steps* video. The original tape of this composition was converted to a QuickTime movie [DM92] which is 1282602 units long. (There are 600 Quick-Time units per second, so this corresponds to a little under 36 minutes.) As can be seen in Figure 12.21, the horizontal dimension of the display accommodates

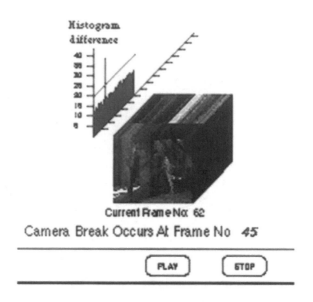

Current Frame No: 62

Camera Break Occurs At Frame No 45

PLAY   STOP

Figure 12.20   A video streamer display.

icons for five frames side by side. Therefore, the whole movie is divided into
five segments of equal length, each of which is represented by the frame at its
midpoint. Thus, for example, the first segment occupies the first 256520 units
of the movie; and its representative frame is at index 128260. Each of these
segments may then be similarly expanded by being divided into five portions
of equal length, each represented by the midpoint frame. By the time we get to
the third level, we are viewing five equally-spaced frames from a segment of size
51304 (approximately 85.5 seconds). (Browsing may be continued to greater
depth, as this material is displayed in a scrollable window.) Furthermore, the
user may select any frame on the display for storage. This will cause the entire
segment represented by the frame to be stored as a separate file which may
then be examined by the micon viewer. (This is how the image in Figure 12.18
was created.)

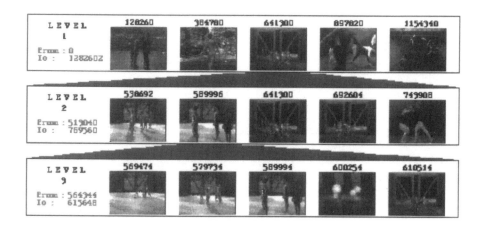

Figure 12.21 A hierarchical browser of a full-length video.

12.6 AUDIO

As any film-maker knows, the audio track provides a very rich source of information to supplement our understanding of any video [BT93]. Thus, any attempt to work with video content must take the sound-track into account as well. As was discussed in Section 12.4.1, logging a video requires decomposing our auditory perceptions into objects just as we do with our visual perceptions. Unfortunately, we currently know far less about the nature of "audio objects" than we do about corresponding visual objects [Smo93]. Nevertheless, there are definitely the beginnings of models of audio events, some of which are similar to models used in image-based content parsing [Haw93]. For example, in a sports video, very loud shouting followed by a long whistle might indicate that someone has scored a goal, which should be recognized by content analysis as an "event." Clearly, however, audio analysis is a new frontier which needs considerable exploration within the discipline of multimedia.

12.7 VIDEO RETRIEVAL IN PRACTICE

Chapter 14 will present a detailed analysis of how the structural analysis of television news may lead to the development of a database of program content. However, before undertaking that discussion, it is worth considering the applications to which video retrieval may be put. Managing video information is a much "younger" domain than image management, as was discussed in Chapter 11. Indeed, it is a domain which has not yet made a significant impression in the marketplace. Nevertheless, technology is developing in such a way that the time for that impression is just about ripe. This Section will discuss three areas of application which are likely to be of increasing significance and importance.

12.7.1   Video-On-Demand

Chapter 3 briefly discussed Video-On-Demand (VOD) as one of the major applications of multimedia technology. VOD promises a future in which individual users will have client devices in their homes which enable them to connect to servers which provide vast quantities and variety of video resources. VOD will eliminate the need to travel to a shop in order to rent a video, let alone handle that video as a physical cassette. Instead, the information will simply exist as digital data transmitted from a server to the user's personal client hardware.

As was observed in Section 12.4.1, however, VOD systems will only be effective to the extent that they can satisfy the "demands," so to speak, of the users. Building servers to hold all those video data and building networks which provide the quality of service discussed in Chapter 2 will only be useful if the user needs the video being provided. How will the user know if the server provides the video he needs? *That* is the problem of retrieval and browsing. Renting a videotape is currently a multimedia experience, enhanced not only by the proprietor's decision to play material on monitors in his shop but also by the visual design of the boxes which hold the videotapes. The home user of a VOD system will require similar multimedia reinforcement as encouragement to seek out video to demand. That encouragement can be provided by either focused query processing which can satisfy specific needs as accurately as can the proprietor of a video rental shop or by less directed browsing, by which the user may be able to sample bits and pieces of available material (perhaps in the form of "previews") before making a final commitment to what he wants. Server and network architectures may provide the data, but retrieval and browsing systems will provide the *need* for those data.

12.7.2   Digital Libraries

VOD is generally viewed as the entertainment of the future. However, we do not always wish to consult video for entertainment purposes. Video material may also be of value for scholarly purposes, such as historical records of events or archives of the output of a creative artist. The need to demand video is just as important for users of libraries of the future as it is for today's video rental customers. Current interest in digital libraries is addressing the need for future access of resources from libraries distributed around the world through computer workstations connected to broadband networks. While many of the resources made available in test studies have been restricted to text, it is clear that digital libraries must also be able to provide access to video [Z+95].

Here, again, both focused query processing and casual browsing will be of importance to future library users. Thus, the same problems of indexing and retrieval which confront VOD systems will also arise in digital libraries. However, library users are more often likely to want to retrieve only *excerpts* of the video they "demand." Thus, the biographer of as ballerina will be interested in all videos of dances in which she performed; but he is likely to want to concentrate on those camera shots in which she appears.

In other words the problem of indexing is similar to that faced by libraries in general. On the one hand there is the "macro" indexing which accounts for all the books and videotapes in the library. However, just as every book has its own index, so must video sources. Having identified a tape of interest, the scholar still needs to extract from that tape those camera shots which are his focus of attention. Unless both these levels of indexes are supported, the user of the digital library will have a hard time getting the most out of his resources.

12.7.3   Film and Television Production

Being able to work with large quantities of video source material is just as important for *creative production* as it is for scholarly research. Thus, individuals who *make* films and television productions are just as concerned with managing, retrieving, and browsing vast video resources as are the users of future digital libraries. There is, however, one distinction which is likely to have technological impact. Scholarship tends to be a solitary pursuit, while the making of a film or video is such a complicated effort that it can seldom be achieved successfully by a single individual [Smo94]. (Only a few rare geniuses, like Charlie Chaplin, stand as exceptions to this general rule of thumb [BT93].) Thus, creative production will require an integration of the functionalities of retrieval and browsing with the technology of computer-supported cooperative work (CSCW) [GCM94].

13

VIDEO PROCESSING USING COMPRESSED DATA

13.1 TECHNIQUES FOR VIDEO PARSING

The processing described in the preceding Chapter is all concerned with operations on pixel representations of individual image frames. However, if a video source is provided in compressed form, these operations cannot be performed until that representation has been decompressed. If decompression can be provided by hardware, this will not constitute a significant computational overhead; but software decompression is far less efficient. If special-purpose hardware is not available, it is worth asking to what extent these operations may be performed directly on compressed representations. We shall now present three difference metrics for frame comparison, two of which employ the DCT coefficients used in both JPEG and MPEG representations, while the third uses MPEG motion vectors. After this, we discuss how to combine the power of the two approaches, incorporating multi-pass, twin-comparison strategies and motion analysis which enable the detection of gradual transitions and camera operations.

It is important to observe that none of these techniques have been deployed in any commercial applications. We are more likely to see a future in which all decompression is relegated to inexpensive hardware than one in which software compression will be a major drain on computational efficiency. On the other hand we have already seen in Chapter 12 that motion vectors provide valuable input for video parsing, and retrieving that input directly from an MPEG representation will be more efficient than any of the methods for computing it discussed in Chapter 7. In other words, because video compression is a signal processing operation, it is capable of deriving certain features which, were they not available to the video parsing process, would have to be re-derived. Thus, even if frame information is readily available through decompression hardware, it is still worth understanding the nature of these features; so they may be exploited for more robust parsing.

13.1.1   Two Algorithms Based on DCT Coefficients

The pioneering work on image processing based directly on compressed data was conducted by Farshid Arman and his colleagues at Siemens Corporate Research [AHC93b]. Their technique consisted in correlating DCT coefficients of consecutive frames of JPEG compressed video. A subset of the DCT coefficients of a subset of the blocks of the frame is extracted to construct a vector representation for that frame:

$$V_f = \{c_1, c_2, c_3, \ldots, c_k\} \tag{13.1}$$

The difference metric between two frames is then defined in terms of a normalized inner product:

$$\Psi = 1 - \frac{|V_f \bullet V_{f+\phi}|}{|V_f||V_{f+\phi}|} \tag{13.2}$$

(ϕ is the number of frames between the two frames being compared.)

An alternative approach is to apply the pair-wise comparison technique discussed in Chapter 12 to the DCT coefficients of corresponding blocks of consecutive video frames [Z+94b]. More specifically, let $c_{l,k}(i)$ be a DCT coefficient of block l in frame i, where k ranges from 1 through 64 (the coefficient number) and l depends on the size of the frame; then the content difference of block l in two frames which are ϕ frames apart can be measured as:

$$Diff_l = \frac{1}{64} \sum_{k=1}^{64} \frac{|c_{l,k}(i) - c_{l,k}(i+\phi)|}{\max[c_{l,k}(i), c_{l,k}(i+\phi)]} \times 100\% \tag{13.3}$$

We can then say that a particular block has changed across the two frames if its difference exceeds a given threshold t:

$$Diff_l > t \tag{13.4}$$

If $D(i, i+\phi)$ is defined to be the percentage of blocks which have changed, then a segment boundary is declared if

$$D(i, i+\phi) > T_b \tag{13.5}$$

where T_b is the threshold for camera breaks. This difference metric is thus analogous to the pair-wise pixel comparison technique of Chapter 12, using DCT coefficients instead of pixel intensities.

Selecting appropriate values for t and T_b may also employ the same techniques discussed in Chapter 12. t tends not to vary across different video sources and can easily be determined experimentally. T_b, on the other hand, is best computed in terms of the overall statistics for values of $D(i, i + \phi)$ as a value which exceeds the mean by about five standard deviations.

For implementation purposes processing time can be reduced significantly by applying Arman's technique of using only a subset of coefficients and blocks. However, the pair-wise comparison algorithm requires far less computation than the difference metric defined by Equation 13.2. Also, it is more sensitive to gradual changes, which makes it more useful than Equation 13.2 for the hybrid approach to detecting transitions which will be discussed in Section 13.1.3.

Both of the above algorithms may be applied to every video frame compressed with Motion JPEG; but in an MPEG video they may only be applied to **I** frames, since those are the only frames encoded with DCT coefficients. Since only a small portion of all video frames are **I** frames, this significantly reduces the amount of processing which goes into computing differences. On the other hand the loss of temporal resolution between **I** frames may introduce false positives which have to be handled with subsequent processing.

The sequences of differences between all successive **I** frames from an MPEG compressed documentary video, defined by both DCT correlation and pair-wise block comparison, are shown in two parts in Figure 13.1 and Figure 13.2, (a) and (b), respectively. This particular sequence contains sharp breaks, gradual transitions, camera operations, and object motion, making it an excellent test case. While it is not much more than a minute long, there are 25 shots separated by 17 sharp breaks and 7 gradual transitions. Also 4 shots involve camera panning. The 17 sharp breaks are labeled 1–17, and they can all be detected by exceeding an appropriately set break threshold T_b. Specific images from this sequence will be presented in the discussion in Section 13.1.3.

However, neither of these two algorithms can handle gradual transitions or false positives introduced by camera operations and object motion. Both Figure 13.1(b) and Figure 13.2(b) illustrate examples of these problems, labeled T_1 through T_7; but we shall not discuss how they may be solved until Section 13.1.3. Instead, we now turn to motion vectors as a source of partitioning information.

13.1.2 Partitioning Based on Motion Vectors

Apart from intensity values and distributions, motion of both objects and the camera are the major elements of video content. In general within a single camera shot, the field of motion vectors should show relatively continuous changes, while this continuity will be disrupted between frames across different shots. Thus, a continuity metric for a sequence of motion vector fields should be able to serve as an alternative criterion for detecting segment boundaries.

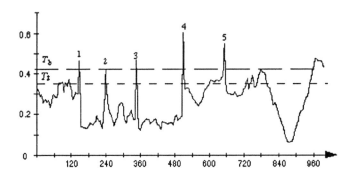

Figure 13.1   (a) Parsing compressed video with difference value Ψ—first half of video compressed in MPEG with 1 intra-picture, 1 predictively coded frame and 4 bi-directional predictively coded frames in every 6 frames: Difference values are computed between successive intra-picture frames. T_b is the threshold for detecting a camera break. T_t is the lower threshold for the twin-comparison approach.

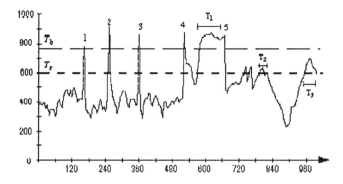

Figure 13.1 *(continued)*
(b) Parsing compressed video with difference value $D(i, i + \phi)$—first half.

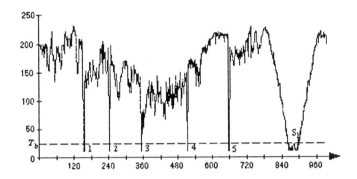

Figure 13.1 *(continued)*
(c) Parsing compressed video with motion vectors—first half.

In an MPEG data stream, as was discussed in Chapter 5, there are two sets of motion vectors associated with each **B** frame, forward and backward, and a single set of motion vectors associated with each **P** frame. That is, each **B** frame is predicted and interpolated from its preceding and succeeding **I/P** frames by motion compensation; and each **P** frame is similarly predicted from its preceding **I/P** frame. In both cases the residual error after motion compensation is then transformed into DCT coefficients and coded. However, if this residual error exceeds a given threshold for certain blocks, motion compensation prediction is abandoned; and those blocks are represented by DCT coefficients, as in **I** frames. Subsequently, there will be no motion vectors associated with those blocks in the **B/P** frames. Such high residual error values are likely to occur

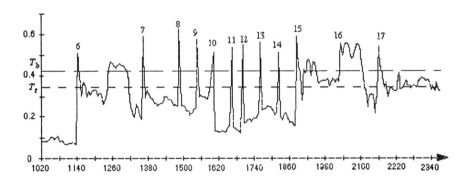

Figure 13.2 (a) Parsing compressed video with difference value Ψ—second half.

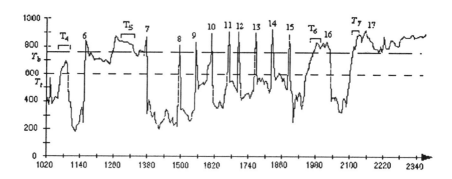

Figure 13.2 *(continued)*
(b) Parsing compressed video with difference value $D(i, i + \phi)$—second half.

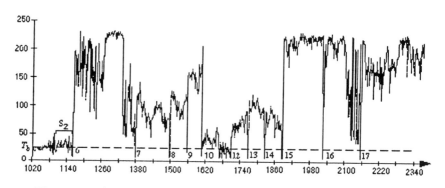

Figure 13.2 *(continued)*
(c) Parsing compressed video with motion vectors—second half.

in many, if not all, blocks across a camera shot boundary; and it is possible to exploit this information.

Let M be the number of valid motion vectors for each **P** frame and the smaller of the numbers of valid forward and backward motion vectors for each **B** frame, and let T_b be a threshold value close to zero. Then

$$M < T_b \tag{13.6}$$

will be an effective indicator of a camera boundary before or after (depending on whether interpolation is forward or backward) the **B/P** frame [Z+94b]. The difference sequences in Figure 13.1(c) and Figure 13.2(c) illustrate these values of M for the same video example used for data in the other graphs in these illustrations. In this case the camera breaks are all accurately represented as valleys below the threshold level, labeled 1 through 17.

However, this algorithm fails to detect gradual transitions because the number of motion vectors during such a transition sequence is often much higher than the threshold. The algorithm also yields false positives, indicated by the S_1 and S_2 valleys. These correspond to sequences of repeating static frames: When there is no motion, all the motion vectors will be zero, yielding a false detection. A simple resolution to this problem is to measure the width of the valley. For a camera break, the valley is usually very deep and narrow, since the small number of motion vectors will be confined to at most two successive frames; on the other hand any static sequence which is longer than two frames will yield a wider valley. However, this may eliminate some gradual transitions with a long sequence of frames with very few motion vectors. The robust solution is to use the difference metric defined in Equation 13.5 to examine the potential false positives, since $D(i, i + \phi)$ will be close to zero between the stationary frames.

13.1.3   A Hybrid Approach to Partitioning

We now present a hybrid approach which integrates the approaches discussed in Sections 13.1.1 and 13.1.2 and incorporates the multi-pass strategies and motion analyses proposed by [Z+94b] to improve detection accuracy and processing speed; we shall base our discussion on video compressed with MPEG.

Multiple Passes and Multiple Comparisons

The first step is to apply a DCT comparison, such as the one defined by Equation 13.5, to the I frames with a large skip factor ϕ to detect regions of potential gradual transitions, breaks, camera operations, or object motion. The large skip factor reduces processing time by comparing fewer frames. Furthermore, gradual transitions are more likely to be detected as potential breaks, since the difference between two more "temporally distant" frames could be larger than the break threshold, T_b. The drawbacks of using a large skip factor, false positives and low temporal resolution for shot boundaries, are then recovered by a second pass with a smaller skip factor (which may be 1, i.e. comparing consecutive I frames); but the second pass is only applied to the neighborhood of the potential breaks and transitions. Thus a high processing speed is achieved without losing either detection accuracy or temporal resolution of segment boundaries.

However, it should be pointed out that choosing a proper skip factor depends on the dynamic features of the movie to be parsed . In the test example discussed in Section 13.2, the rapid changes due to moving objects and camera motion make a large skip factor relatively inadequate. Since, in general, the frame interval between consecutive I frames in MPEG movies is larger than six frames, an even larger skip factor will not increase the overall processing speed because of the number of false positives introduced.

What makes this a hybrid approach is that the motion-based comparison metric is also applied as another pass, either on the entire video or only on the sequences containing potential breaks and transitions , to complement and verify the results from DCT comparison and to improve further the accuracy and confidence of the detection results. More specifically, the difference metric of Equation 13.6 is applied to all the B and P frames of those selected sequences to confirm and refine the break points and transitions detected by the DCT comparison. Conversely, the DCT results provide information for distinguishing between sequences of static frames and transition frames which could be confused by motion-based detection as observed in Section 13.1.2. A good example of an advantage of this combined approach is break 2 in Figure 13.1(a), which is missed by the DCT correlation algorithm but is clearly recognized from motion vector comparison. The images compared across this break (as well as those across break 3 for control) are illustrated in Figure 13.3.

Gradual transitions may be detected by an adaptation of the twin-comparison approach discussed in Chapter 12. For example T_4 in Figure 13.2(b) can be correctly detected as a gradual transition (illustrated in Figure 13.4), since the DCT differences between I frames exceed the lower threshold, T_t, of the twin-comparison approach. On the other hand T_1 in Figure 13.1(b) is a camera pan (illustrated in Figure 13.5) whose difference values are higher than the break threshold, due to the large frame interval between I frames (6 in our case). However, because motion vector analysis is very robust in detecting sharp breaks, this false positive can be detected. Therefore, by fusion of the information gathered from different passes with different comparison metrics, the false positives and missing transitions, resulting from using any of the three metrics alone, may be resolved.

Figure 13.3 Transitions for breaks 2 and 3 as indicated in the graphs. Break 2 is illustrated in the upper pair of images. Break 3 is illustrated in the lower. Break 3 was detected by DCT coefficient correlation, while break 2 was missed but was detected by motion vector comparison.

Camera Operation and Object Motion Detection in the Hybrid Approach

Another video parsing problem discussed in Chapter 12 consists in distinguishing sequences involving camera operations from those due to gradual transitions, since the former tend to induce temporal variations in frame content of the same order as do the latter. The solution presented in Chapter 12 is based on detecting specific patterns in the field of motion vectors. While that analysis included the problem of computing those vectors, the same technique may be applied to the vectors which are provided by MPEG **B** frames, yielding virtually identical results.

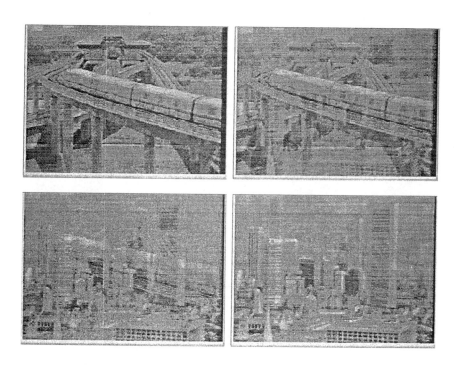

Figure 13.4   A gradual transition corresponding to segment T_4.

13.2   EXPERIMENTAL DATA

To give a quantitative evaluation of the different algorithms presented above, we summarize the results from Figure 13.1 and Figure 13.2 in tabular form. First let us consider the detection of camera breaks. In Table 13.1 DCT1 refers to the DCT correlation comparison, defined in Equation 13.2, while DCT2 is the pair-wise block comparison technique. It is observed that both the hybrid and motion vector algorithms detected all breaks correctly, while the two DCT algorithms produced a number of false positives. DCT2 also missed a break. Thus, for break detection in MPEG compressed video, the motion vector data gave the best performance in terms of both speed and accuracy. The false detection in DCT algorithms mainly resulted from the fact that only **I** frames can been used in these algorithms. Due to the large interval between consecutive **I** frames and the effects of motion, DCT-based algorithms are much less effective when they are applied to MPEG data streams than they are with motion JPEG, where every frame is DCT-coded and thus can be compared.

Figure 13.5   A camera pan corresponding to segment T_1.

	Detected	Undetected	Falsely Detected
DCT1	16	1	4
DCT2	17	0	4
motion vector	17	0	0
hybrid	17	0	0

Table 13.1   Summary of results in break detection.

The detection of gradual transitions is similarly summarized in Table 13.2. Note that only the hybrid algorithm detected all transitions correctly. The two DCT-based algorithms missed a few and produced one false positive, and the motion vector analysis failed as was discussed in Section 13.1.2. Thus, for the highest accuracy in detecting both camera breaks and gradual transitions, the hybrid approach is the optimal choice. Finally, the camera operation detection algorithm described in Section 13.1.3 successfully detected all four of the camera pans in the test data.

	Detected	Undetected	Falsely Detected
DCT1	3	5	1
DCT2	4	4	1
motion vector	0	7	0
hybrid	7	0	0

Table 13.2 Summary of results in gradual transition detection.

14

A CASE STUDY IN VIDEO
PARSING: TELEVISION NEWS

14.1   INTRODUCTION

Automatic extraction of "semantic" information of general video programs is
outside the capability of current machine vision and audio signal analysis technologies. On the other hand "content parsing" may be possible when one has
an *a priori* model of a video's structure based on domain knowledge. Such a
model may represent a strong spatial order within the individual images and/or
a strong temporal order across a sequence of shots. A television news program
is a good example of a video which follows such a structural model: there tends
to be spatial structure within the anchorperson shots and temporal structure
in the order of shots and episodes.

The temporal syntax of a news video is usually very straightforward. Figure
14.1 shows an example of the episode structure of a news program: It is a simple
sequence of news items (possibly interleaved with commercials), each of which
may include an anchorperson shot at its beginning and/or end. Individual
news shots are, as a rule, not easily classified by syntactic properties, with the
possible exception of certain regular features, such as weather, sports, and/or
business. Parsing thus relies on classifying each camera shot according to these
relatively coarse categories.

We consider such syntactic analysis to be the first yet important step towards
video content understanding, which will enable the construction of indexes to
facilitate content-based video retrieval. This first step automatically extracts
two basic index units of a news program (news items and camera shots), clas-
sifies each item into a coarse category, and extracts key frames to represent the
visual content of each news item. Such syntactic information can support the

335

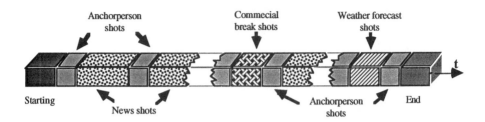

Figure 14.1   The temporal structure of a typical news program.

technology of video indexing techniques in three ways: First, both indexing and data entry benefit from automatic provision of the following information: classification of individual camera shots, identification of consecutive camera shots in a common segment, counting of shots, identification of the starting time and the duration of each shot (and segment), and, most important, extraction of a set of key frames which represent the visual content of each shot. Second, syntactic processing improves the productivity in current human logging processes by allowing the user to concentrate on recording information that cannot be provided automatically. Third, and perhaps most important, the syntactic information will facilitate subsequent semantic analysis of certain segments, such as sports news, since only after such segments have been located can they be analyzed by more specific domain knowledge, such as recognizing a goal being scored in a football game.

When we consider retrieval , rather than indexing, we see other benefits of syntactic analysis. For example, our syntactic categories are too coarse to be able to distinguish an international news segment from a local news segment. However, the extraction of key frames will almost always provide sufficient visual cues to enable the user to make this more refined classification on the basis of a display of the results. Furthermore, because many anchor desk shots are accompanied by an iconic image which serves as a "headline," that icon may be extracted and enlarged, to be used as an additional semantic cue to content. Thus, what cannot be analyzed automatically can be presented in such a way as to facilitate manual analysis.

This syntactic approach was proposed by Deborah Swanberg and her colleagues [SSJ93] as a technique for managing video databases. However, no experimental results were reported. A more recent study of knowledge-guided video content parsing algorithms has led to the development and implementation of a system for automatic news program parsing and indexing [Z+94a]. This Chapter uses the results of that project to treat the domain of television news as a case study for more general work on video parsing, indexing , retrieval , and browsing. The most important elements of this case study are its approaches to selecting the models which drive the parsing process, identifying anchorperson shots, and indexing and representing a resulting parsed news video.

14.2 NEWS VIDEO PARSING ALGORITHMS

The video parsing process of [Z+94a] consists of three steps: temporal segmentation to partition a video sequence into shots, shot classification based on the spatial structure of an anchorperson shot model and a temporal structure model of an entire news program, and, finally, visual abstraction to extract a set of key frames from each shot to represent the visual content of each news item. Temporal segmentation has already been discussed in Chapter 12, so this Chapter will concentrate on algorithms and methods for shot classification and visual abstraction. Of greatest importance is the structure of a typical anchorperson shot, which we shall call an "**A** shot" for simplicity, which usually consists of a sequence of frames containing an anchorperson with a picture news icon in the upper left or right part of the frames. Because news programs often use two anchorpersons, the shots may also include a sequence of frames containing one or two anchorpersons with a "label bar" of their names and, sometimes, the title of the program. In contrast the news shots tend not to have any fixed temporal and spatial structure. Thus, to a first approximation, news shots may be identified as those that do not conform to the anchorperson model.

14.2.1 Shot Classification

Parsing a news program may thus be broken down into three tasks: The first task is to define a model of an **A** shot which incorporates both the temporal structure of the shot and the spatial structure of a representative frame. The second task is to develop similarity measures to be used in matching these models with a given shot as a means of deciding whether that shot is an **A** shot. The third task is to use a temporal structure model of the entire news program to finalize the shot classification. These processes will now be discussed in detail.

Data Modeling

The first task in shot classification is data modeling. [SSJ93] proposed constructing three models for an anchorperson shot: shot, frame, and region. An anchorperson shot is modeled as a sequence of frame models, and a frame is modeled as a spatial arrangement of region models. Thus, recognizing an **A** shot involves testing that every frame satisfies a frame model, which, in turn, means testing each frame against a set of region models. In addition a temporal model is applied to the entire news program which also includes the starting and ending sequences of frames of the program.

Frame Models Figure 14.2 illustrates the spatial structure of a typical anchorperson shot in a news program of the Singapore Broadcasting Corporation (SBC);[1] and Figure 14.3 shows four models with different spatial structures. The possibilities for variation lie in the number of anchorpersons, the presence or absence of a news icon, and the presence or absence of labeling bars. In general, if there is no news icon in a single **A** frame, the anchorperson will be in the center. Otherwise, the anchorperson will be either on the left or the right side, with the news icon on the opposite side. In a two-anchorperson frame there is usually no news icon.

Figure 14.2   The spatial structure of a typical anchorperson shot.

---

[1] Now the Television Corporation of Singapore

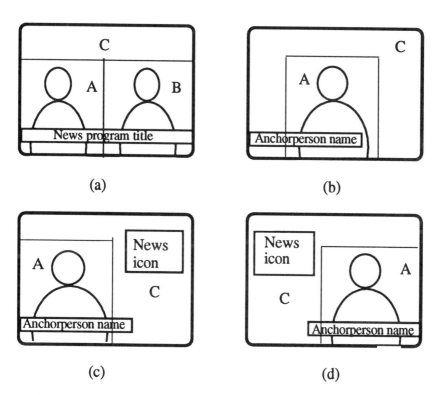

Figure 14.3 Spatial structure models of anchorperson shots: (a) two anchorpersons; (b) one anchorperson; (c) one anchorperson with an upper-right news icon; (d) one anchorperson with an upper-left news icon.

Region Models The frame models listed in Figure 14.3 are based on the following region models: anchorperson, news icon, news program title bar, anchorperson name bar, and background. Identifying an **A** frame requires first classification of its regions according to these models. Variations across news programs broadcast at different times or even by different networks may be reflected in variations in both the frame and region models.

Anchorperson Shots Between News Items An entire **A** shot consists of a sequence of frames conforming to the models in Figure 14.3. If the background region tends to be relatively fixed, the color histogram of a representative frame may be included as part of the frame model. The temporal feature which most effectively models **A** shots is the magnitude of frame-to-frame changes (Chapter

12). Since these tend to be static talking-head shots, this magnitude is generally low.

Starting and Ending Sequences   Special sequences, very different from the **A** shots, are usually employed at the beginning and end of a news program, and sometimes when the news continues after commercial breaks. There is great variation in these sequences across different productions which, most recently, have reflected a great infatuation with special effects which provide very rapid content change. Nonetheless, any given production will employ these sequences very consistently, using them as its "signature" across all its broadcasts. Temporal sequences of frame-to-frame differences, where the differences may be computed from either pixel intensities or their histograms, provide satisfactory models of these special segments.

Weather Forecast Shots   There is considerable variation here. On the one hand SBC conforms to a very simple model involving the display of text and icons against a uniform background. *CNN Headline News*, on the other hand, displays a map that exhibits different markings for every broadcast (reflecting the changes in meteorological conditions). In general, however, the frame which is displayed tends to be identical across the entire shot.

News Shots   There is too much variety in news shots to support any viable structure model. On the other hand the frame-to-frame differences between consecutive frames in news shots are generally larger than those between anchorperson frames, since there is much more action and use of camera operations. Such differences are especially evident when the pair-wise pixel comparison metric (Chapter 12) is used.

Episode Model   This is basically the model illustrated in Figure 14.1. There is a starting sequence followed by a sequence of anchorperson shots during which the highlights are reviewed. There is then an alternation between anchorperson shots and news shots, usually one cycle for each news item, with perhaps commercial breaks inserted. The structure of some programs, such as *CNN Headline News*, may be further refined to consist of sub-episodes for special topics such as finance and sports.

Detecting Potential **A** Shots

Given a segmented video, the task here is to identify those segments that are **A** shots. Swanberg has taken a template matching approach, where the templates are typically other images, called *model images*; and matching is based on either a weighted difference measure or a weighted similarity measure [SSJ93].

Unfortunately, the amount of variation over a set of similarly produced news programs may be too high for a fixed set of *a priori* model images to be effective. Variations in both who the anchorpersons are and what they are wearing from one program to the next will probably make matching against a single model very difficult, if not impossible. Furthermore, the matching algorithm itself can be very time-consuming, given the number of shots to which it may have to be applied. Therefore, an alternative to using predefined model images is an approach which first locates *potential* **A** shots using temporal features. Model images can then be acquired from these candidates for model matching.

As we stated previously, the average difference, by any metric, between two consecutive frames of an **A** shot will be generally smaller than that between consecutive frames of news shots, since the change in the former is introduced only by minor body movement and random noise. Therefore, for a given metric, we can calculate the mean, μ, and variance, σ^2, of the difference values over the entire shot. That shot can then be declared a potential **A** shot if these statistics satisfy the following simple threshold inequalities:

$$\mu \quad < t_1 \qquad\qquad (14.1a)$$
$$\sigma^2 \quad < t_2 \qquad\qquad (14.1b)$$

These candidates can then be further verified by model matching.

It is actually desirable to use statistics based on both the histogram *and* the pair-wise pixel comparison metrics of Chapter 12 to achieve highest accuracy in locating potential **A** shots. This is because, on the one hand, the histogram metric is less sensitive to motion between frames; but it may be too insensitive to distinguish **A** shots from some news shots, resulting in missing **A** shot candidates. On the other hand the pair-wise pixel comparison metric may be too sensitive: minor body movement by the anchorperson may be falsely interpreted as large change. Therefore, the two metrics are used to complement each other. Only those shots within which the μ and σ^2 statistics for *both* metrics satisfy their respective thresholds should be classified as potential **A** shots.

Model Image Selection

We next take a candidate **A** shot and try to detect the appearance of anchorperson(s). This involves searching for variations within the regions characterized by the region models shown in Figure 14.3. For example, in a shot of two anchorpersons, changes between two consecutive frames will only occur in the areas of the two bodies (primarily in the faces). Therefore, we divide the frames into regions, A, B and C, as shown in Figure 14.3, and calculate the mean and variance of the differences in these regions between consecutive frames of the candidate shots. Since there are head movements in regions A and B but no motion or change in region C, we have

$$\mu_A > \mu_t > 0 \qquad\qquad \sigma_A^2 > \sigma_t^2 > 0 \qquad\qquad (14.2a)$$

$$\mu_B > \mu_t > 0 \qquad\qquad \sigma_B^2 > \sigma_t^2 > 0 \qquad\qquad (14.2b)$$

$$\mu_C \approx 0 \qquad\qquad\qquad \sigma_C \approx 0 \qquad\qquad\qquad (14.2c)$$

Thus, a shot can be declared as an **A** shot if the above conditions are satisfied. We can also use this technique to establish which of the models shown in Figure 14.3 is the applicable one.

After the shot is confirmed to be an **A** shot, a model image is obtained as follows: we calculate the average frame of the shot by defining each pixel as:

$$p_A(i,j) = \frac{1}{N} \sum_{n=1}^{N} p_n(i,j) \qquad\qquad (14.3)$$

In this formula N is the number of frames inside the shot, and $p_n(i,j)$ is the value of a pixel at frame n. Then the average image $\{p_A(i,j)\}$ is stored as a model image for the corresponding frame model and will be used to identify the same type of **A** shots among the remaining candidates. Actually, it is only necessary to model the anchorperson regions A and B, as defined in Figure 14.3, for matching. This not only reduces the computation time (fewer pixels are calculated) but also avoids a problem of variation in the rest of the frame (induced, for example, by changing news icons).

A *Shot Matching Process*

In matching a candidate frame to the model image, the absolute difference between the corresponding regions is used as the similarity measure:

$$S = 1 - \frac{1}{M \times N} \sum_{i,j}^{M,N} |p(i,j) - p_A(i,j)| \qquad\qquad (14.4)$$

Each model image of a region also includes its intensity histogram. Thus, before computing the similarity measure, if the histogram of a candidate shot does not match the model's histogram, it will be immediately rejected, providing further reduction in processing time.

Model images are constructed sequentially from the beginning of the broadcast. Once a candidate **A** shot is verified as a given type, the region image(s) and histogram(s) are stored as the model for that type. If a new candidate shot is of a new type, then the model construction process is repeated for that type. In this way the models are constructed incrementally until all types have been modeled. After this point, each new candidate is compared with all available models. Detection error may be tolerated by setting the maximum number of shot models to be more than the actual number of frame types. Then, models with no match in the subsequent processes will be discarded. This will eliminate mistaking an interview news shot for an **A** shot because of its similar temporal variation and spatial structure .

Context information is also used in this classification process. If the two shots before and after a shot to be classified are both **A** shots, then this shot will most likely be an **A** shot also. This is due to the fact that between two news shot sequences there may be two or three **A** shots in a sequence to bridge two news items. Another constraint is that between two news sequences there are usually not more than three different anchorperson shots: an **A** shot ending the previous news, an **A** shot starting the next news without a news icon, and finally an **A** shot with a news icon. There may also be an opportunity to exploit stylistic constraints imposed on the length of news shots.

14.2.2   Miscellaneous Sequence and Episode Identification

Since the starting and ending sequences of a news program have a fixed and pre-defined temporal and spatial structure, they may be easily identified from the data stream. Similarly, the pre-defined temporal and spatial structure of the sequence used when the program returns from a commercial break can be used to detect the commercial breaks. Figure 14.4 shows four frames of a starting sequence of an SBC news program. Therefore, for a given news program, we can classify all shots into the following categories: anchorperson shots; news shots; commercial break shots; weather forecast shots; and starting and ending shots. Finally, the episodes are identified by matching a model of the entire news program, as shown in Figure 14.1, to the sequence of classified shots.

Figure 14.4 Four frames from the starting sequence on an SBC news program.

14.2.3   Visual Abstraction: Key Frames

Once each shot in every news item has been identified, it may be abstracted by a set of representative images, usually called *key frames* . This is a familiar technique in manually logging a video. Fortunately, the process may be automated as was discussed in Chapter 12.

14.3   EXPERIMENTAL EVALUATION OF PARSING ALGORITHMS

We shall now review the evaluation of the techniques summarized in Section 14.2 using, as test data, two half-hour SBC news programs. Figure 14.5 shows the flow diagram of the parsing process. When the system parses the data, it decompresses the data into single frames and performs the temporal feature analysis and spatial structure matching processes described in Section 14.2. Also, a list of camera shot boundaries is obtained using the segmentation techniques described in Chapter 12.

Figure 14.6, together with Figure 14.2, shows examples of **A** frames corresponding to the four models shown in Figure 14.3. Testing has demonstrated that these types of anchorperson shots are correctly identified by the techniques discussed in Section 14.2. First they are recognized as potential **A** shots. Then the anchorperson regions are correctly selected as regional model images using the temporal variation conditions defined by Equation 14.2a.

The starting and ending sequences, as well as the return from commercial breaks, of the two programs have been correctly identified using both the timing information and the predefined frame and sequence models shown in Figure 14.4. Given this episode information, the news program can then be segmented

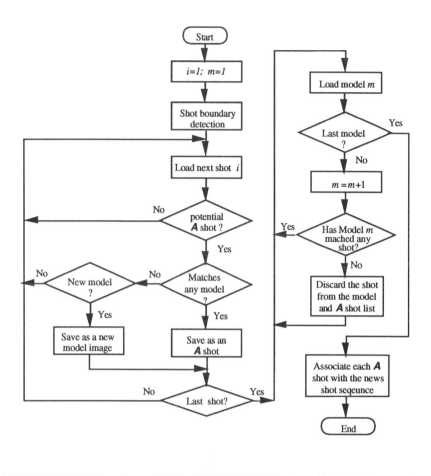

Figure 14.6 Anchorperson frames from SBC news programs: (left) two anchorpersons; (center) one anchorperson; (right) one anchorperson with upperleft news icon.

into a sequence of news shots, **A** shots, commercial breaks, weather shots, and starting and ending sub-sequences, as defined in the episode model shown in Figure 14.1.

Figure 14.7 Examples of shots detected as potential anchorperson shots, but eliminated in the parsing process: (left) eliminated in the model image selection process; (right) eliminated in the matching process.

Program	N_{total}	N_p	N_A	N_{Af}	N_{Am}
SBC1	549	92	23	0	1
SBC2	409	80	31	3	0

Table 14.1 Experiment results in detecting anchorperson shots: N_{total}, total number of shots in the news program; N_p, number of potential anchorperson shots identified by the temporal variation analysis; N_A, number of anchorperson shots finally identified; N_{Af}, false positives; N_{Am}, missed anchorperson shots; SBC1, first half-hour broadcast of SBC news; SBC2, second half-hour broadcast of SBC news.

Figure 14.7 shows two examples of frames from the shots that have been determined as potential **A** shots due to the low temporal variations between the frames of the shots. The left-hand image is from a sequence consisting of graphics and text frames, between which the temporal change is very low. In the sequence from which the right-hand image is taken, the camera focuses on a speaker, resulting in similarly small temporal variation. Examples of the first type are eliminated by the model image selection process, since there is virtually no temporal variation in *any* region of the frames. Examples of the second type are eliminated by the model matching process, since, for each instance of such a shot, the regional model image matches no other shot in the list but itself.

Table 14.1 lists the numbers of video shots and anchorperson shots identified by the system. Over 95% of the **A** shots were detected correctly. The missing anchorperson shot in SBC1 was due to the fact that, in that shot, the anchorperson was reading a news item without facing the camera. As a result, the shot does not match any anchorperson models extracted from the program. There are a few false positives in both programs, where an interview scene was repeated in several shots and was falsely detected as an **A** shot.

Program	N	N_S	N_m	N_f
SBC1	20	18	2	0
SBC2	19	18	1	0

Table 14.2 Experiment results in detecting news items: N, number of news items manually identified by watching the programs; N_S, news items identified by the system; N_m, news items missed by the system; N_f, news items falsely identified by the system.

Table 14.2 lists the numbers of news items identified by the system and the numbers manually identified by watching the programs. It is seen that the system has identified the news items with a very high accuracy (greater than 95%), which shows that the algorithms are effective and accurate. The false positives of **A** shots did not introduce false detection of news items in the second program. This is because news items tend to contain two **A** shots with the same background, so this condition will not be confused by a falsely identified **A** shot. The missed news items resulted from the assumption that each news item in the program starts with an **A** shot followed by a sequence of news shots. However, this condition can be violated in the two news programs by both news items which are only read by an anchorperson without news shots and by news items which start *without* an **A** shot. The latter case is responsible for the missed news items in the experiment, when the news item started with only the anchorperson's voice in the background. Another possibility is that two news items share a single **A** shot. This is the major limitation of the approach currently being used, and it is difficult to overcome with only image analysis techniques. In such cases content analysis requires taking the audio signal into account [Haw93] as well.

14.4 NEWS VIDEO INDEXING AND RETRIEVAL

The ultimate objective in parsing news video programs is to provide indexes to their content which will organize the storage of these programs in a database. The basic index unit of a news program in this case is the news item, allowing a user to access directly any such news item of interest, rather than watching the news sequentially. Thus, it is essential to provide the user a set of fast retrieval and browsing tools for accessing and viewing the video shots classified by the parsing process.

The first task for indexing is to organize the news items identified by parsing for easy access. The index information for each such news item includes the number of shots in the news item, the starting time, the duration, and, most importantly, a set of key frames which represent the visual content of each shot. Apart from these attributes, derived automatically by the news parser, text descriptions about the news content can be input by an operator. Two index schemes can accommodate these attributes: a text index and a visual index. The news database can be designed in an object-oriented manner.

Following the usual conventions of a book index, the text index is based on a tree of topical categories [Z+95], where news items are assigned to leaves of the tree, just as page numbers are assigned to entries in a book index. Each node of the tree then corresponds to a collection of news items that have some amount of topical information in common. The information common to all those items is stored as a class frame in the object-oriented representation. Figure 14.8 [Z+95] shows a portion of a category tree for one of the SBC news programs. Note that, at the news program level, categories have been defined to correspond to the syntactic units of the temporal structure illustrated in Figure 14.1; but the actual content of the news items has been refined to represent different types of news stories.

Once a hierarchy of categories has been defined, it can also be used to facilitate any logging or annotation processes. Each category will have associated with it some characteristic set of fields that delimit the information to be recorded during logging. Many of these fields are filled in *automatically* with results from the parsing process. Consequently, logging becomes a far less tedious and more focused activity. Routine information is provided automatically, and the logger need only record material that is characteristically unique to each log entry.

If the category tree is displayed in a window as a hierarchy of "hot spots," that window can be used as a browsing and retrieval tool such that a user can browse down from a news program to a particular news item, and to a particular shot within the news item. In contrast, the tree can be reconstructed such that all news items of the same type, but broadcast by various programs at various times, are collected in one category and can be examined as a single group. This is especially useful when users do not know beforehand what they are searching for and often will need to *recognize* what they are looking for rather than *describe* it.

Apart from category-based retrieval, text retrieval techniques may be applied to key word and free-text descriptions of news items. Given a text key, the system first extracts the relevant terms by removing the nonfunctional words

```
NewsProgram
      Transition
            start
            end
      NewsContent
            Summary
            International
                  news1
                  news2
                  news3
                  news4
                  news5
                  news6
                  news7
                  news8
                  news9
            Local
                  news10
                  news11
            Financial
                  business_today
                  financial_summary
```

Figure 14.8   A portion of a topic category tree of news video programs.

and stemming the remainder. Then, the query is checked against a domain-specific thesaurus, after which similarity measures are used to compare the text descriptions with the query terms. If the similarity measure of news items or programs exceeds a given threshold, they are identified and retrieved, then linearly ordered by the strength of the similarity [ZS94].

The retrieved news items or shots of a query can be displayed as a list of names or visually presented as a *clip window* [ZS94], as shown in Figure 14.9. Figure 14.9 shows all the news items from a single program which have been assigned to the "International" class in Figure 14.8. Each news item is represented by an icon of a rectangular solid the "depth" of which represents its duration. The front face of the icon is called its *cover frame* and provides a representative image of the entire contents. Where possible, this cover frame is an anchorperson frame containing a news icon, which provides a visual cue to the entire news story. If the opening **A** shot does not contain a news icon, then the cover frame will be the first frame of the first news shot following that **A** shot.

A clip window not only displays the results of a retrieval but also serves as an interface for more detailed examination of the video source. Clicking a news icon brings up another clip window, in which each icon represents a single camera shot from that news item. (An example of one of these clip windows is shown in Figure 14.10.) This allows direct access to each shot. In addition, at either level of temporal resolution, an icon may be loaded into a "soft" video player; and its entire contents may be viewed. This visual approach to the examination of news items and shots also makes retrieval a more *interactive* process than that of simply formulating and processing conventional queries. This leads to fewer constraints on the possibilities the user may wish to explore.

The actual database of video content is organized around the key frames of each shot in each news item. This means that it is not necessary to view or store full resolution video data in a database representation of content. Key frames are extracted according to the process discussed in Chapter 12. Once they are obtained, their visual features are derived, quantified and stored as part of the index to the database using the techniques discussed in Chapter 11.

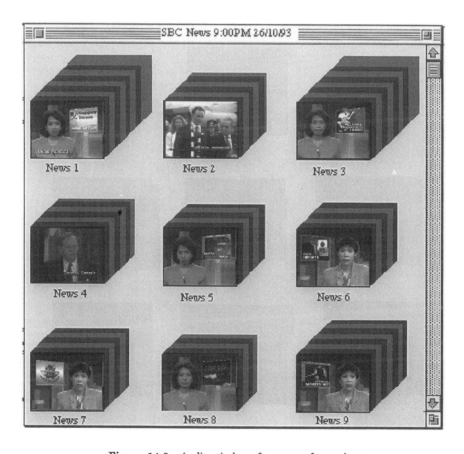

Figure 14.9 A clip window of a group of news items.

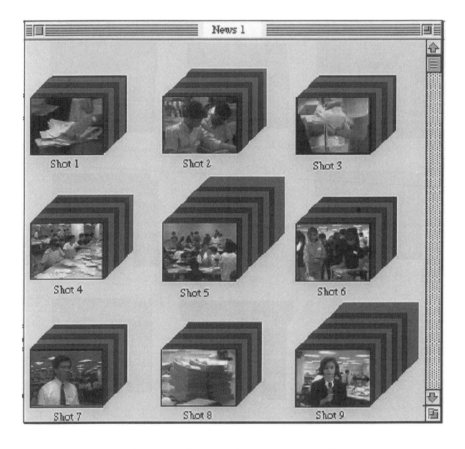

Figure 14.10   Presenting the shots of a news item in a clip window.

14.5 MODELING OTHER DOMAINS OF CONTENT

14.5.1 Football

In retrospect it is not surprising that news program material is relatively easy to parse. Its temporal and spatial predictability tend to make it easier for viewers to pick up those items which are of greatest interest. (Predictability in the layout of a daily newspaper serves a similar function.) However, because more general video program material tends to be concerned with entertainment, which, in turn, is a matter of seizing and holding viewers' attention, the success of a program often hinges on the right balance between the predictable and the *un*predictable. Thus, the very elements which often make a program successful are those which would confound attempts to automate parsing.

However, not all video material which requires parsing is prepared for broadcast for entertainment. Video can also be collected as an information resource; and if the material on such a video is to be effectively catalogued for a database, then it must first be relatively systematically parsed. This parsing (or annotation or logging) process is generally performed manually and is a tedious and error-prone technique. However, if a body of video material can be suitably modeled, much of that process may be automated, making it a more efficient and possibly more reliable task.

An example of such a resource is a database of football plays [Int94]. At Boston College these plays are recorded with a Betacam situated at the top of the stadium on the fifty-yard line (center of the playing field). The camera has an operator who makes all decisions to pan and zoom, depending on the nature of the play being recorded.

In this case parsing is less a matter of temporal segmentation. The camera operator only records game plays; so there are well-defined breaks between each segment. The important element of *content* in a play is the *movement* of all the players, the referee, and the football in the course of a play. Thus, the parsing problem is one of tracking all such movement and presenting the results in a useful format.

General object tracking remains a difficult problem. Many different approaches to its solution have been proposed, each of which has its own characteristic set of limitations [Int94]. However, in this case assumptions about where the camera is and what it sees greatly simplifies the problem. One can begin with a model of the entire football field and rectify every image with a transformation which maps it to the appropriate subset of that model. Through the model it is then possible to identify certain regular visual features, most of which are the grid lines which mark the field in a standardized manner. (Because all videos were collected at Boston College, it was also possible to account for a logo in the center of the field.) Given all these predictable features, detecting movement becomes a more feasible task. Results can then be presented as trace lines on the model image of the entire football field and may even be embellished to use the sorts of representations commonly used to illustrate football plays [Int94]. Such static images which represent the action in an entire play then significantly enhance the content of a database of those plays.

14.5.2   Dance

A similar resource of information recorded on video is dance. Both film and video are frequently used to record both rehearsals and performances; and (unfortunately rarely) performances are recorded under studio conditions and subsequently edited for broadcast or commercial distribution. Often this may be the only information available to anyone interested in recreating that particular dance. A variety of notation systems have been developed (by way of analogy to music notation); but most dancers and choreographers are "notation illiterates." The notation can only be used if a specialist is on hand, both to record in the notation and to physically demonstrate what the notation has recorded [BS79].

A growing interest in digital libraries has led to a related interest in the creation of video databases for scholarly and reference purposes [Z$^+$95]. As a result interest is also growing in incorporating existing libraries of film and video of dance in this future vision. This raises the question of whether or not dance databases are as feasible as those for football plays. Much of the visual content of a dance may actually be captured reasonably effectively using key frame extraction techniques, such as were discussed in Chapter 12. However, such key frames are static images; they communicate little about the motion content which is the actual dance.

Unfortunately, there is far too much variation in the nature of the movement, the background, and the camera work to accommodate a model similar to that discussed in Section 14.5.1. On the other hand the ability to construct micons, such as were described in Chapter 12, for each camera shot of a dance

Figure 14.11 A spatiotemporal image illustrating traces of dancers' legs (©1994 IEEE).

provide another approach to representing movement information. Figure 14.11 [SZ94] is an example of a micon constructed from a portion of the *Changing Steps* video used for several of the examples in Chapter 12. Following the approach discussed in Chapter 12, this micon may be cut with a horizontal slice at approximately knee height, constructing a spatiotemporal image which provides a trace of each dancer's legs [Ade91]. The micon may then be rectified into an "extruded solid" which accounts for camera movement, also discussed in Chapter 12; so the traces then provide a representation of movement within a fixed frame of reference. They may then be compared against floor-plan representations of dancer movement which are frequently documented [BS79] using the sketch-based retrieval techniques discussed in Chapter 11. What is important, then, is that movement content may be captured by suitably constructed static images.

The greatest shortcoming with this approach is that it assumes that all feet are always on the ground, which is not always the case in dance. However, a model-based approach may be used to supplement this technique to establish, for any given camera shot, where it is most appropriate to construct one or more spatiotemporal images. Indeed, the number of cases in which alternative cuts across a micon are required will probably be quite few, since dancers seldom leave the ground for very long. More important is that the model provide a heuristic for establishing knee height, regardless of the orientation of the camera.

REFERENCES

[A+92]   A. Akutsu et al. Video indexing using motion vectors. In *Visual Communications and Image Processing '92*, pages 1522–1530, Boston, MA, November 1992. SPIE.

[A+94]   F. Arman et al. Content-based browsing of video sequences. In *Proceedings: ACM Multimedia 94*, San Francisco, CA, October 1994. ACM.

[AB85]   E. H. Adelson and J. R. Bergen. Spatiotemporal energy models for the perception of motion. *Journal of the Optical Society of America A*, 2(2):284–299, February 1985.

[Ade91] E. Adelson. Mechanisms for motion perception. *Optics and Photonics News*, 2(8):24–30, August 1991.

[AH+91] S. Al-Hawamdeh et al. Nearest neighbour searching in a picture archive system. In *International Conference on Multimedia Information Systems '91*, pages 17–33, SINGAPORE, January 1991. ACM, McGraw Hill.

[AHC93a] F. Arman, A. Hsu, and M.-Y. Chiu. Feature management for large video databases. In W. Niblack, editor, *Symposium on Electronic Imaging Science and Technology: Storage and Retrieval for Image Video Databases*, pages 2–12, San Jose, CA, February 1993. IS&T/SPIE.

[AHC93b] F. Arman, A. Hsu, and M.-Y. Chiu. Image processing on compressed data for large video databases. In *Proceedings: ACM Multimedia 93*, pages 267–272, Anaheim, CA, August 1993. ACM.

[ANAH93] Y.-H. Ang, A. D. Narasimhalu, and S. Al-Hawamdeh. Image information retrieval systems. In C. H. Chen, L. F. Pau, and P. S. P. Wang, editors, *Handbook of Pattern Recognition and Computer Vision*, chapter 4.2, pages 719–739. World Scientific, SINGAPORE, 1993.

[AS92]   T. G. Aguierre Smith. If you could see what I mean...descriptions
           of video in an anthropologist's video notebook. Master's thesis,
           Massachusetts Institute of Technology, Cambridge, MA, September
           1992.

[AW93]   E. H. Adelson and J. Y. A. Wang. Representing moving images
           with layers. Technical Report 228, MIT Media Lab Perceptual
           Computing Group, Cambridge, MA, April 1993.

[BC92]   N. J. Belkin and W. B. Croft. Information filtering and information
           retrieval: Two sides of the same coin? *Communications of the
           ACM*, 35(12):29–38, December 1992.

[BF81]   A. Barr and E. A. Feigenbaum, editors. *The Handbook of Artificial
           Intelligence*, volume 1, chapter 3, pages 141–222. William Kauf-
           mann, Los Altos, CA, 1981.

[Bro66]   P. Brodatz. *Textures: A Photographic Album for Artists and De-
           signers*. Dover, New York, 1966.

[BS79]   N. I. Badler and S. W. Smoliar. Digital representations of human
           movement. *Computing Surveys*, 11(1):19–38, March 1979.

[BT93]   D. Bordwell and K. Thompson. *Film Art: An Introduction*. Mc-
           Graw Hill, New York, NY, fourth edition, 1993.

[Caw93]   A. E. Cawkill. The British Library's picture research projects: Im-
           age, word, and retrieval. *Advanced Imaging*, 8(10):38–40, October
           1993.

[CF82]   P. R. Cohen and E. A. Feigenbaum, editors. *The Handbook of
           Artificial Intelligence*, volume 3, chapter 13, pages 125–321. William
           Kaufmann, Los Altos, CA, 1982.

[CH92]   S. K. Chang and A. Hsu. Image information systems: Where do
           we go from here? *IEEE Transactions on Knowledge and Data
           Engineering*, 4(5):431–442, October 1992.

[Cha89]   S. K. Chang. *Principles of Pictorial Information Systems Design*.
           Prentice-Hall, Englewood Cliffs, NJ, 1989.

[Dat77]   C. J. Date. *An Introduction to Database Systems*. The Systems
           Programming Series. Addison-Wesley, Reading, MA, second edi-
           tion, 1977.

[Dav93]   M. Davis. Media streams: An iconic visual language for video an-
           notation. In *Proceedings: Symposium on Visual Languages*, pages
           196–202, Bergen, NORWAY, 1993. IEEE.

[DG94]   N. Dimitrova and F. Golshani. R_x for semantic video database
           retrieval. In *Proceedings: ACM Multimedia 94*, San Francisco, CA,
           October 1994. ACM.

[DH73]   R. Duda and P. Hart. *Pattern Classification and Scene Analysis.*
           Wiley, New York, NY, 1973.

[DM92]   D. L. Drucker and M. D. Murie. *QuickTime Handbook.* Hayden,
           Carmel, IN, 1992.

[Ede87]   G. M. Edelman. *Neural Darwinism: The Theory of Neuronal Group
           Selection.* Basic Books, New York, NY, 1987.

[Ede89]   G. M. Edelman. *The Remembered Present: A Biological Theory of
           Consciousness.* Basic Books, New York, NY, 1989.

[Ell93]   E. L. Elliott. Watch • grab • arrange • see. Master's thesis, Mas-
           sachusetts Institute of Technology, Cambridge, MA, February 1993.

[F+94]   C. Faloutsos et al. Efficient and effective querying by image content.
           Journal of Intelligent Information Systems, 3:231–262, 1994.

[FA92]   W. T. Freeman and E. H. Adelson. The design and use of steer-
           able filters. *IEEE Transactions on Pattern Analysis and Machine
           Intelligence*, 38(2):587–607, 1992.

[G+94]   Y. Gong et al. An image database system with content capturing
           and fast image indexing abilities. In *Proceedings of the International
           Conference on Multimedia Computing and Systems*, pages 121–130,
           Boston, MA, May 1994. IEEE.

[GCM94]   E. Gidney, A. Chandler, and G. McFarlane. CSCW for film and
           TV preproduction. *IEEE MultiMedia*, 1(2):16–26, Summer 1994.

[Gib86]   J. J. Gibson. *The Ecological Approach to Visual Perception.* Erl-
           baum, Hillsdale, NJ, 1986.

[GWJ91]   A. Gupta, T. Weymouth, and R. Jain. Semantic queries with pic-
           tures: The VIMSYS model. In *Proceedings of the 17th Interna-
           tional Conference on Very Large Databases*, pages 69–79, Barcelona,
           SPAIN, September 1991.

[GZ94]   Y. H. Gong and H. J. Zhang. An effective method for detecting
           regions of given colors and the features of the region surfaces. In
           S. A. Rajala and R. L. Stevenson, editors, *Symposium on Electronic
           Imaging Science and Technology: Image and Video Processing II*,
           pages 274–285, San Jose, CA, February 1994. IS&T/SPIE.

[Haw93]   M. J. Hawley. *Structure out of Sound*. PhD thesis, Massachusetts
           Institute of Technology, Cambridge, MA, September 1993.

[Hei77]   M. Heidegger. Being and time: Introduction. In D. F. Krell, editor,
           Basic Writings from Being and Time *(1927) to* The Task of Think-
           ing *(1964)*, chapter 1, pages 37–89. HarperCollins, New York, NY,
           1977. Translated from the German by J. Stambaugh in collabora-
           tion with J. G. Gray and D. F. Krell.

[HJW94]   A. Hampapur, R. Jain, and T. Weymouth. Digital video segmen-
           tation. In *Proceedings: ACM Multimedia 94*, San Francisco, CA,
           October 1994. ACM.

[HS81]   B. K. P. Horn and B. G. Schunck. Determining optical flow. *Arti-
           ficial Intelligence*, 17:185–203, 1981.

[Hu77]   M. K. Hu. Visual pattern recognition by moment invariants. In
           J. K. Aggarwal, R. O. Duda, and A. Rosenfeld, editors, *Computer
           Methods in Image Analysis*. IEEE Computer Society, Los Angeles,
           CA, 1977.

[Hun89]   L. E. Hunter. Knowledge acquisition planning: Gaining expertise
           through experience. Technical Report YALEU/DCS/TR-678, Yale
           University, New Haven, CT, January 1989.

[Hus70]   E. Husserl. *The Crisis of European Sciences and Transcenden-
           tal Phenomenology*. Northwestern University Press, Evanston, IL,
           1970. Translated from the German, with an Introduction, by D.
           Carr.

[Int94]   S. S. Intille. Tracking using a local closed-world assumption: Track-
           ing in the football domain. Technical Report 296, MIT Media Lab
           Perceptual Computing Group, Cambridge, MA, August 1994.

[Iok89]   M. Ioka. A method of defining the similarity of images on the
           basis of color information. Technical Report RT-0030, IBM Tokyo
           Research Laboratory, Tokyo, JAPAN, November 1989.

[Jai89]   A. K. Jain. *Fundamentals of Digital Image Processing*. Prentice-
           Hall, Englewood Cliffs, NJ, 1989.

[K+92]   T. Kato et al. A sketch retrieval method for full color image database: Query by visual example. In *Proceedings: 11th International Conference on Pattern Recognition*, pages 530–533, Amsterdam, HOLLAND, September 1992. IAPR, IEEE.

[KJ91]   R. Kasturi and R. Jain. Dynamic vision. In R. Kasturi and R. Jain, editors, *Computer Vision: Principles*, pages 469–480. IEEE Computer Society Press, Washington, DC, 1991.

[KK87]   A. Khotanzad and R. L. Kashyap. Feature selection for texture recognition based on image synthesis. *IEEE Transactions on Systems, Man, and Cybernetics*, 17(6):1087–1095, November 1987.

[Koh90]   T. Kohonen. The Self-Organizing Map. *Proceedings of the IEEE*, 78(9):1464–1480, September 1990.

[KZL94]   A. Kankanhalli, H. J. Zhang, and C. Y. Low. Using texture for image retrieval. In *Third International Conference on Automation, Robotics and Computer Vision*, pages 935–939, SINGAPORE, November 1994.

[LG89]   D. B. Lenat and R. V. Guha. *Building Large Knowledge-Based Systems: Representation and Inference in the Cyc Project*. Addison-Wesley, Reading, MA, 1989.

[LP94]   F. Liu and R. W. Picard. Periodicity, directionality, and randomness: Wold features for perceptual pattern recognition. In *Proceedings: 12th International Conference on Pattern Recognition*, pages 184–189, Jerusalem, ISRAEL, October 1994. IAPR, IEEE. Volume II.

[MCW92]   M. Mills, J. Cohen, and Y. Y. Wong. A magnifier tool for video data. In *Proceedings: CHI'92*, pages 93–98, Monterey, CA, May 1992. ACM.

[MJ92]   J. Mao and A. K. Jain. Texture classification and segmentation using multiresolution simultaneous autoregressive models. *Pattern Recognition*, 25(2):173–188, 1992.

[MS81]   D. McLeod and J. M. Smith. Abstraction in databases. *SIGPLAN Notices*, 16(1):19–23, January 1981. Also SIGART Newsletter, Number 74, and SIGMOD Record, Volume 11, Number 2.

[N+93]   W. Niblack et al. The QBIC project: Querying images by content using color, texture and shape. In *Symposium on Electronic Imaging*

Science and Technology: Storage and Retrieval for Image Video Databases, San Jose, CA, February 1993. IS&T/SPIE.

[NP83] R. M. Nowak and J. L. Paradiso. *Walker's Mammals of the World*. The Johns Hopkins University Press, Baltimore, MD, fourth edition, 1983.

[NT92] A. Nagasaka and Y. Tanaka. Automatic video indexing and full-video search for object appearances. In E. Knuth and L. M. Wegner, editors, *Visual Database Systems, II*, volume A-7 of *IFIP Transactions A: Computer Science and Technology*, pages 113–127. North-Holland, Amsterdam, THE NETHERLANDS, 1992.

[O'C91] B. C. O'Connor. Selecting key frames of moving image documents: A digital environment for analysis and navigation. *Microcomputers for Information Management*, 8(2):119–133, June 1991.

[PG93] R. W. Picard and M. Gorkani. Finding perceptually dominant orientations in natural textures. Technical Report 229, MIT Media Laboratory Perceptual Computing Group, Cambridge, MA, 1993.

[PKL93] R. W. Picard, T. Kabir, and F. Liu. Real-time recognition with the entire Brodatz texture database. In *Proceedings: IEEE Conference on Computer Vision and Image Processing*, pages 638–639, New York, NY, June 1993. IEEE.

[PM94] R. W. Picard and T. P. Minka. Vision texture for annotation. Technical Report 302, MIT Media Laboratory Perceptual Computing Group, Cambridge, MA, 1994.

[PPS94] A. Pentland, R. W. Picard, and S. Sclaroff. Photobook: Tools for content-based manipulation of image databases. In W. Niblack and R. Jain, editors, *Symposium on Electronic Imaging Science and Technology: Storage and Retrieval for Image Video Databases II*, pages 34–47, San Jose, CA, February 1994. IS&T/SPIE.

[Pra91] W. K. Pratt. *Digital Image Processing*. Wiley, New York, NY, second edition, 1991.

[RBE94] L. A. Rowe, J. S. Boreczky, and C. A. Eads. Indexes for user access to large video databases. In W. Niblack and R. C. Jain, editors, *Symposium on Electronic Imaging Science and Technology: Storage and Retrieval for Image Video Databases II*, pages 150–161, San Jose, CA, February 1994. IS&T/SPIE.

[RHM86]   D. E. Rumelhart, G. E. Hinton, and J. L. McClelland. A general framework for parallel distributed processing. In *Parallel Distributed Processing: Explorations in the Microstructure of Cognition*, volume 1, chapter 2, pages 45–76. The MIT Press, Cambridge, MA, 1986.

[SB91]   M. J. Swain and D. H. Ballard. Color indexing. *International Journal of Computer Vision*, 7(1):11–32, 1991.

[See92]   A. N. Seeley. *User Guide: Aldus Fetch Version 1.0*. Aldus Corporation, Seattle, WA, first edition, November 1992.

[Sey94]   IBM unleashes QBIC image-content search. *Seybold Report on Desktop Publishing*, 9(1), September 1994.

[SFP94]   R. Sriram, J. M. Francos, and W. A. Pearlman. Texture coding using a Wold decomposition model. In *Proceedings: 12th International Conference on Pattern Recognition*, pages 35–39, Jerusalem, ISRAEL, October 1994. IAPR, IEEE. Volume III.

[SM83]   G. Salton and M. McGill. *Introduction to Modern Information Retrieval*. McGraw-Hill, New York, NY, 1983.

[Smo93]   S. W. Smoliar. Classifying everyday sounds in video annotation. In T.-S. Chua and T. L. Kunii, editors, *Multimedia Modeling*, pages 309–313, SINGAPORE, November 1993.

[Smo94]   S. W. Smoliar. On the promises of multimedia authoring. *Information and Software Technology*, 36(4):243–245, April 1994.

[SSJ93]   D. Swanberg, C.-F. Shu, and R. Jain. Knowledge guided parsing in video databases. In *Symposium on Electronic Imaging: Science and Technology*, San Jose, CA, 1993. IS&T/SPIE.

[SZ94]   S. W. Smoliar and H. J. Zhang. Content-based video indexing and retrieval. *IEEE MultiMedia*, 1(2):62–72, Summer 1994.

[SZW94]   S. W. Smoliar, H. J. Zhang, and J. H. Wu. Using frame technology to manage video. In *Second Singapore International Conference on Intelligent Systems*, pages B189–B194, SINGAPORE, November 1994.

[T+93]   Y. Tonomura et al. VideoMAP and VideoSpaceIcon: Tools for anatomizing video content. In *Proceedings: INTERCHI '93*, pages 131–136, 544, Amsterdam, NETHERLANDS, April 1993. ACM.

[TB93]   L. Teodosio and W. Bender. Salient video stills: Content and con-
            text preserved. In *Proceedings: ACM Multimedia 93*, pages 39–46,
            Anaheim, CA, August 1993. ACM.

[TC92]   D. C. Tseng and C. H. Chang. Color segmentation using perceptual
            attributes. In *Proceedings: 11th International Conference on Pat-
            tern Recognition*, pages 228–231, Amsterdam, HOLLAND, Septem-
            ber 1992. IAPR, IEEE.

[TJ93]   M. Tuceryan and A. K. Jain. Texture analysis. In C. H. Chen, L. F.
            Pau, and P. S. P. Wang, editors, *Handbook of Pattern Recognition
            and Computer Vision*, chapter 4.2, pages 235–276. World Scientific,
            SINGAPORE, 1993.

[TMY76]   H. Tamura, S. Mori, and T. Yamawaki. Texture features corre-
            sponding to visual perception. *IEEE Transactions on Systems,
            Man, and Cybernetics*, 6(4):460–473, April 1976.

[Ton91]   Y. Tonomura. Video handling based on structured information for
            hypermedia systems. In *International Conference on Multimedia
            Information Systems '91*, pages 333–344, SINGAPORE, January
            1991. ACM, McGraw Hill.

[W+94]   J. K. Wu et al. Inference and retrieval of facial images. *Multimedia
            Systems*, 2(1):1–14, 1994.

[Wit74]   L. Wittgenstein. *Philosophical Investigations*. Basil Blackwell, Ox-
            ford, England, 1974. Translated by G. E. M. Anscombe.

[Z+94a]   H. J. Zhang et al. Automatic parsing of news video. In *Proceed-
            ings of the International Conference on Multimedia Computing and
            Systems*, pages 45–54, Boston, MA, May 1994. IEEE.

[Z+94b]   H. J. Zhang et al. Video parsing using compressed data. In
            W. Niblack and R. Jain, editors, *Symposium on Electronic Imag-
            ing Science and Technology: Image and Video Processing II*, pages
            142–149, San Jose, CA, February 1994. IS&T/SPIE.

[Z+95]   H. J. Zhang et al. A video database system for digital libraries. In
            N. R. Adam, B. Bhargava, and Y. Yesha, editors, *Advances in Dig-
            ital Libraries*, Lecture Notes in Computer Science. Springer Verlag,
            Berlin, GERMANY, 1995. To appear.

[ZKS93]   H. J. Zhang, A. Kankanhalli, and S. W. Smoliar. Automatic parti-
            tioning of full-motion video. *Multimedia Systems*, 1(1):10–28, 1993.

[ZLS95]   H. J. Zhang, C. Y. Low, and S. W. Smoliar. Video parsing and browsing using compressed data. *Multimedia Tools and Applications*, 1(1):91–113, February 1995.

[ZS94]   H. J. Zhang and S. W. Smoliar. Developing power tools for video indexing and retrieval. In W. Niblack and R. Jain, editors, *Symposium on Electronic Imaging Science and Technology: Storage and Retrieval for Image Video Databases II*, pages 140–149, San Jose, CA, February 1994. IS&T/SPIE.

[ZZ95]   H. J. Zhang and D. Zhong. Scheme for visual feature-based image indexing. In W. Niblack and R. Jain, editors, *Symposium on Electronic Imaging Science and Technology: Storage and Retrieval for Image Video Databases III*, San Jose, CA, February 1995. IS&T/SPIE.

QUESTIONS AND PROBLEMS - PART III

1. Let $I(i)$ be an N-bin normalized histogram of an image, where $I(i)$ is the function which maps bin i to its population. We can define the following features over any histogram:

Moments: $m_k = E[i^k] = \sum_{i=1}^{N} i^k I(i)$

Absolute moments: $\hat{m}_k = E[|i^k|] = \sum_{i=1}^{N} |i|^k I(i)$

Central moments: $\mu_k = E\{[i - E(i)]^k\} = \sum_{i=1}^{N} (i - m_1)^k I(i)$

Absolute central moments: $\hat{\mu}_k = E[|i - E(i)|^k] = \sum_{i=1}^{N} |i - m_1|^k I(i)$

Try to use the following metrics for image matching and compare the results with those based on histogram intersection:

- dispersion $= \hat{\mu}_1$

- mean $= m_1$

- variance $= \mu_2$

- average energy $= m_2$

- skewness $= \mu_3$

- kurtosis $= \mu_4 - 3$

2. The orientation of an object is defined to be the angle of the axis of its least moment of inertia. Derive Equation 11.24 based on this definition and the central moments defined in Equation 11.20.

3. The *best fit ellipse* of an object is the ellipse whose second-order moment equals that of the object. This is illustrated as follows:

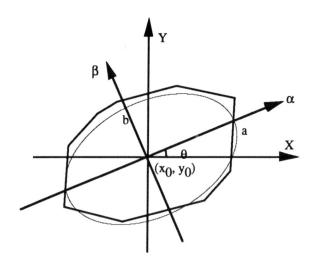

Let a and b denote the lengths of the semimajor and semiminor axes of the best fit ellipse, respectively. Then the least and greatest moments of inertia for the ellipse are defined as follows:

$$I_{\min} = \frac{\pi}{4}ab^3, \qquad I_{\max} = \frac{\pi}{4}a^3b$$

Use these equations and the definitions of moments to derive values for a and b of the best fit ellipse.

4. Use the theory of wavelet-based image compression (discussed in Chapter 8) to design an algorithm to represent texture features which may be applied for texture-based image retrieval.

5. Use the ISODATA clustering algorithm (described in *Pattern Classification and Scene Analysis* by R. Duda and P. Hart) to build an index tree based on the color histograms of a set of images.

6. Apply the definition of the Tamura directionality feature (Chapter 11) to spatiotemporal images as a means of detecting dominant motion directions which may be used as cues for camera operations.

7. Design a scheme for retrieving camera shots based on extracted key frames. What sorts of queries may be supported by this scheme?

8. Design a similarity metric based on histogram data from an entire camera shot.

9. How may the data in a wavelet-based compression of a video be applied to solve the temporal segmentation problem?

10. How may the data in a fractal-based compression of a video be applied to solve the temporal segmentation problem?

11. Choose a source of news broadcasts. How is it necessary to adapt the spatial modesl of anchorperson shots to accommodate this program material? Design a matching algorithm to locate instances of these models in a program.

12. Adapt the techniques discussed in Chapter 14 to work on a motion JPEG compression of a news program.

13. Design a parsing model (temporal and spatial) for a popular form of commercial television programming, such as a basketball game broadcast.

INDEX